Advances in Experimental Medicine and Biology

Proteomics, Metabolomics, Interactomics and Systems Biology

Series Editor

Daniel Martins-de-Souza
University of Campinas (UNICAMP)
Institute of Biology
Laboratory of Neuroproteomics
Campinas, Brazil

Advances in Experimental Medicine and Biology presents multidisciplinary and dynamic findings in the broad fields of experimental medicine and biology. The wide variety in topics it presents offers readers multiple perspectives on a variety of disciplines including neuroscience, microbiology, immunology, biochemistry, biomedical engineering and cancer research. *Advances in Experimental Medicine and Biology* has been publishing exceptional works in the field for over 30 years and is indexed in Medline, Scopus, EMBASE, BIOSIS, Biological Abstracts, CSA, Biological Sciences and Living Resources (ASFA-1), and Biological Sciences. The series also provides scientists with up-to-date information on emerging topics and techniques. 2018 Impact Factor: 1.760

More information about this series at http://www.springer.com/series/15040

José-Luis Capelo-Martínez

Editor

Emerging Sample Treatments in Proteomics

 Springer

Editor
José-Luis Capelo-Martínez
BIOSCOPE Group, LAQV-REQUIMTE
Departamento de Química
Faculdade de Ciências e Tecnologia
Universidade Nova de Lisboa
Caparica, Portugal

Proteomass Scientific Society
Rua dos Inventores
Madan Park
Caparica, Portugal

ISSN 0065-2598   ISSN 2214-8019   (electronic)
Advances in Experimental Medicine and Biology
ISBN 978-3-030-12300-0   ISBN 978-3-030-12298-0   (eBook)
https://doi.org/10.1007/978-3-030-12298-0

This Springer imprint is published by the registered company Springer Nature Switzerland AG
The registered company address is: Gewerbestrasse 11, 6330 Cham, Switzerland

Preface

The word "proteome" was coined in 1994 by Marc Wilkins to define a group of proteins encoded by a genome and expressed in particular cells, tissues, organs, or organisms. A quarter of a century later, proteomics, the science that studies the proteome, has become one of the most important and popular scientific fields for the large-scale study of complex cell-based systems, largely boosted by the parallel development of affordable high-resolution mass spectrometers and powerful bioinformatics tools. As a current example, the hyphenation of ion-mobility and time-of-flight-based mass spectrometry with parallel accumulation-serial fragmentation software has made it possible to quantify about 1500 proteins in 20 minutes, a staggering throughput at an incredible speed [1].

Nowadays, the only limit to the wide range of proteomics applications is our imagination. The study of biological functions, understanding changes in cellular regulation mechanisms caused by disease states, discovery of biomarkers for disease diagnosis, prognosis, and the development of new drugs for therapeutic approaches can all benefit from the use of proteomics. Just 10 years ago, most proteomics studies relied on different but complementary tools, such as two-dimensional gel electrophoresis, chromatographic separation, and mass spectrometry. Protein quantification was, and in many cases still is, done using stable isotope labeling methods. Protein quantification by label-free shot-gun ion-mobility high-resolution mass spectrometry is soon likely to supersede isotope-based methods. Software for data collection and analysis has also greatly evolved, moving from data-dependent to data-independent acquisition.

However, despite such technological advances, the procedures used to handle samples in the laboratory are still complex, lengthy, and laborious. Sample collection and preservation and proteome extraction, purification, reduction, alkylation, and digestion are key steps that greatly influence the pipeline of mass spectrometry-based quantification, especially the reliability of the results. One of the most important steps in any protein identification or quantitation workflow is the digestion, or hydrolysis, of the proteome. This crucial step is traditionally performed with proteases, such as trypsin, for 12–48 hours. With the advent of the so-called ProteoSonic method, the digestion of 96 proteomes can be performed in less than 5 minutes, an unprecedented speed of sample throughput [2].

The improvement of two critical steps remains to be tackled before protein quantification, using mass spectrometry, can be considered a routine technique. One is the development of software to handle the vast quantity of information generated with new ion mobility high-resolution mass spectrometers. The other, sample treatment, is the subject of this book. There is no universal treatment to extract the proteome from all types of samples. This is because the matrixes of cells from different tissues and organisms present different constraints due to their specific physical and chemical properties, so different approaches are required to successfully extract as many of the proteins as possible. For this reason, after a first chapter addressing the necessary tools of the trade of modern proteomics, eight chapters follow, focusing on different sources of biological samples, namely, saliva, plasma, serum, animal tissues, urine, feces, plants, and bacteria. Each chapter describes in detail the current sample treatments used daily by researchers in the analytics and proteomics communities, including the pitfalls and how to overcome them, and the pros and cons to consider.

I would like to thank Professor Daniel Martins-de-Souza, who kindly encouraged me to edit this book, and the Springer Editorial team, who kindly accepted my book proposal. I send my deepest gratitude to all chapter contributors and thank them for their valuable insights and shared interest in this vibrant field.

Caparica, Grande Lisboa, Portugal   José-Luis Capelo-Martínez

References

1. timsTOF Pro – The new standard for shotgun proteomics. https://www.bruker.com/products/mass-spectrometry-and-separations/lc-ms/o-tof/timstof-pro/overview.html. Last accessed 24 March 2019
2. Jorge S, Araújo JE, Pimentel-Santos FM, Branco JC, Santos HM, Lodeiro C, Capelo JL (2018) Unparalleled sample treatment throughput for proteomics workflows relying on ultrasonic energy. Talanta 178:1067–1076. https://doi.org/10.1016/j.talanta.2017.07.079

Contents

Chapter 1
Proteomics: Tools of the Trade

Utpal Bose, Gene Wijffels, Crispin A. Howitt, and Michelle L. Colgrave

1.1 Introduction

Proteins are the functional traits of life, and they play critical roles in orchestrating various biological processes in order to define the cellular or organismal phenotype. Collectively the protein complement of a cell, organ or organism is known as the proteome. The proteome describes a dynamic system where the presence and abundance of proteins facilitate a multitude of processes and regulatory mechanisms. Individual proteins can carry out functions in living cells, or they can interact to form higher order structures and/or networks that work synergistically within a complex system [1]. The structure and function of proteins are altered by post-translational modifications, protein-protein interactions, their subcellular location and as a result of synthesis and degradation the progression of biological processes, such as cell proliferation, migration and development [2]. In order to understand the underlying mechanism of physiological processes, it is critical to be able to identify and quantify proteins and their interactions within these dynamic systems.

Sample preparation is an integral and crucial part of any proteomics experiment. The primary goal of sample preparation is to obtain high-quality protein samples which are representative of the proteome. Thus sample preparation is critical to maximising protein identification and quantitation rates in downstream analyses. Biological samples are complex, and they contain significant amounts of non-protein components that can interfere with protein analyses. To extract, identify, quantify and characterise proteins from these samples, diverse protein extraction

U. Bose · G. Wijffels · M. L. Colgrave (✉)
Agriculture and Food, CSIRO, St Lucia, QLD, Australia
e-mail: Utpal.bose@csiro.au; gene.wijffels@csiro.au; michelle.colgrave@csiro.au

C. A. Howitt
Agriculture and Food, CSIRO, Canberra, ACT, Australia
e-mail: crispin.howitt@csiro.au

© Springer Nature Switzerland AG 2019
J.-L. Capelo-Martínez (ed.), *Emerging Sample Treatments in Proteomics*, Advances in Experimental Medicine and Biology 1073, https://doi.org/10.1007/978-3-030-12298-0_1

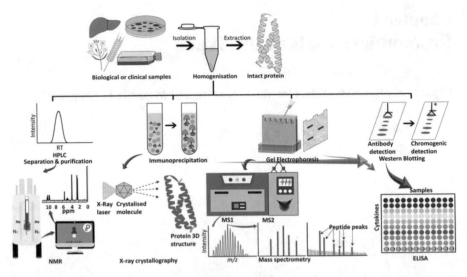

Fig. 1.1 Sample preparation for antibody-based techniques (e.g. immunoprecipitation, gel electrophoresis and Western blotting), structural proteomics (e.g. NMR and X-ray crystallography) and top-down MS (e.g. MALDI)

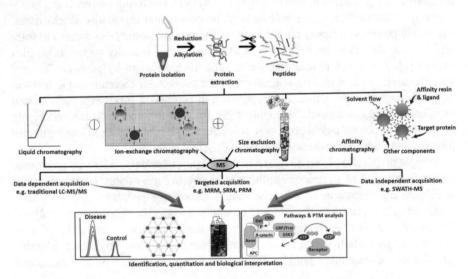

Fig. 1.2 Steps involved in MS-based proteomics sample preparation and data analysis

and purification techniques have been used to prepare samples for the different types of analytical methods. Proteomics experiments typically comprise at least four steps: (1) sample collection and storage (e.g. sample collection from biological or clinical specimen and storage at required temperature); (2) protein extraction (e.g. reagent-based or mechanical disruption); (3) protein clean-up or purification (e.g. chromatography, protein precipitation, enrichment or depletion); and (4) protein analyses (e.g. antibody-based techniques, mass spectrometry, structural studies). The optimal sample preparation method will be highly dependent on the analytical goal. For instance, sample preparation for liquid chromatography-mass spectrometry (LC-MS) proteomics will be vastly different to that used for nuclear magnetic resonance (NMR) structure elucidation experiments [3]. Less sample is required for LC-MS-based bottom-up proteomics, but the workflow will typically include a protein digestion step prior to identification or quantitation of the proteins; whereas a substantially higher amount of highly purified target protein is required for NMR-based structural elucidation experiments.

Proteins play critical roles in various cellular processes, and their complete characterisation is of great interest. However, it is technically challenging to determine the presence, abundance, modification state(s) and localisation of every protein in a system due to the complex environment in which they exist. An array of well-established methods and protocols including immunohistochemistry (IHC), immunoprecipitation (IP), Western blotting (WB) or enzyme-linked immunosorbent assay (ELISA), in vitro protein assays and in vivo imaging are used to study the presence, sequence, function and biochemical mode of action of one or a few specific proteins. With techniques such as mass spectrometry (MS), it is now possible to list the cast of protein players and using systems biology approaches start to elucidate the cellular and/or organismal priorities in a particular state or time point [1]. For example, LC-MS proteomics quantified ~11,000 proteins from human heart [4] or yeast (*Saccharomyces cerevisiae*) [5] within a few hours.

The ideal protein extraction and identification method should capture the most comprehensive repertoire of proteins, and ideally their abundances, while minimising protein degradation and/or the effect of non-protein interferences. This chapter will focus on sample pretreatment for proteomics experiments as it applies to commonly used methods and instruments used in proteomics experiments in the context of their principles and experiment-specific applications (Figs. 1.1 and 1.2). The overall considerations in proteomic sample preparation are given in Box 1.1.

Box 1.1 Considerations for the Protein Sample Preparation Strategy
- *Aim:* What is the ultimate goal of the project, identification, characterisation or quantitation?
- *Type:* What is the sample type? Eukaryotes, prokaryotes, plants, etc., vary in their cell/tissue structure requiring different approaches.
- *Scale:* How much sample do you have? Can you cover technical and biological replicates? Concentration or enrichment steps might be required. Selection of a sensitive method of analysis may overcome low sample amounts.
- *Localisation:* Where is/are the protein/s of interest localised? Reagent and protocol selection will define the success in isolating protein(s) from a

Box 1.1 (continued)

subcellular fraction or organelle. Subcellular fractionation can assist in enriching low-abundant components. Harsher solvents or strong detergents are often required to efficiently extract intrinsic membrane proteins.

- *Dynamic range:* Is/are the protein(s) of interest in a complex matrix? Is the aim to extract all proteins or a particular target? The detection and/or quantitation of low-abundance target proteins may require depletion, enrichment or purification steps and increase the amount of sample required due to inevitable losses and degradation.
- *Preservation:* Is protein structure or function important? Inhibitors (e.g. protease or phosphatase) may be added to preserve the proteins. Should thermal denaturation of enzymes be considered to limit post-collection changes?
- *Analysis platform:* What type of downstream analytical platform or methodologies will be used? Gels, WB, ELISA, MS, NMR, X-ray crystallography. The level of purity is dictated by the downstream analyses. For example, detergents and salts interfere with MS and need to be removed. Highly purified protein is often required for protein structure identification by NMR or X-ray crystallography.

1.2 Steps Involved in Proteomics Sample Preparation

1.2.1 Sample Collection

Sample collection from biological or clinical sources is the first step in proteomic sample preparation. Samples can be obtained from naïve source material such as whole tissues, body fluids or cell culture. Samples representing all kingdoms—animal, plants, fungi, or from the prokaryotic sources such as bacteria—are commonly investigated with each sample type presenting unique challenges in their analysis. During sample collection, the cells or tissues are removed from their origin, which can cause disturbances in homeostasis. This step can trigger the activity of various proteases and enzymes that are responsible for post-translational modifications (PTMs) leading to alteration or degradation of protein components. Numerous strategies have been used to reduce the chance of protein loss or modification. These include rapid freezing in liquid nitrogen, chemical- and/or temperature-induced denaturation of protein-modifying enzymes and the addition of protective or stabilising compounds such as reducing agents of protease inhibitors. Protein stabilisation may also achieved by thermal denaturation or protein precipitation. Should samples need to be preserved for future experiments, tissue samples or extracted proteins can be stored at −80 °C. However, the rapid extraction of proteins from frozen samples is preferred as some samples may be less stable than others.

1.2.2 Protein Extraction

Protein sample extraction from biological or clinical sources can profoundly affect the experimental outcome. Generally, protein extraction methods can be divided into two main categories: gentle processes, for instance, chemical and osmotic lysis, liquid or liquid/liquid extraction and temperature treatments, and harsher processes, such as mechanical disruption via ultrasonication, grinding and/or pressure homogenisation. In chemical-based lysis processes, denaturing solutions (lysis buffers) are often used to disturb protein structural elements, leading to disruption of protein-protein interaction and protein unfolding and thus facilitating protein extraction. Strong denaturants (e.g. urea, guanidine, heat, detergents) and extremes of pH and ionic detergents (e.g. sodium dodecyl sulphate (SDS), deoxycholate and cetyl-trimethylammonium bromide) are often used to solubilise and denature membrane proteins. Non-ionic or zwitterionic detergents (e.g. Triton X-100, NP-40, digitonin or CHAPS) have a lower micelle concentration and thus require a lower amount to solubilise proteins [6]. In cells and prokaryotes, lysozymes (enzyme) can be used to lyse erythrocytes and for the extraction of periplasmic proteins from E. coli by subcellular fractionation [7]. In addition to chemical-based approaches, liquid or liquid/liquid protein extraction methods have been included in the workflow to remove cell debris and as a preliminary clean-up step in proteomic sample preparation [8]. An appropriate combination of organic solvents can be selected to extract proteins based on their phase separation or partitioning. The extraction efficiency can be improved by altering the pH and temperature, varying the ionic strength and salt type and changing polymer or surfactant concentrations.

Although chemical-based extraction is often the first choice, mechanical or physical disruption methods are particularly useful in protein extraction from plant tissues. The most common methods in this category include grinding under liquid nitrogen (mostly for plant tissues), use of ultrasonic probes, lysis by osmotic shock, grinding with aluminium or sand, pressure cycling, use of bead beaters/devices and thermal lysis [9, 10]. Mechanical disruption methods can suffer from poor reproducibility in part resulting from non-homogeneity of the finished sample after processing. To mitigate this issue, chemical/solvent-based extraction is often used in combination with homogenisation and mechanical grinding to efficiently extract proteins from diverse sample types.

1.2.3 Protein Depletion

Biological and clinical samples represent an excellent opportunity to discover biomarkers for their future application as diagnostics or as therapeutic targets. However, it can be challenging to identify and consistently quantify low-abundance biomarker proteins in the presence of high-abundance proteins [11]. For example, human plasma and serum protein concentrations in the blood can span up to 12 orders of magnitude [12]. They contain a substantial amount of albumin and immunoglobulin that can comprise up to 70% of the protein content [11]. Different protein depletion strategies, such as protein precipitation and antibody column-based approaches, have been developed to mitigate this problem, where the ultimate goal is to separate high-abundance proteins from the low-abundance proteins. Protein depletion methods, such as albumin precipitation [13, 14] or column-based immunodepletion [15], and commercial kits such as ProteoPrep® 20 (Sigma-Aldrich), Pierce™ Top (Thermo Fisher Scientific) and Aurum Affi-Gel Blue Mini Columns (Bio-Rad) have been developed to substantially reduce the high abundance proteins from blood serum and plasma in order to identify low-abundance proteins.

In plants, ribulose-1,5-bisphosphate carboxylase oxygenase (Rubisco) represents 30–50% of the total protein in green tissues, which imposes a consistent challenge for both gel-based and shotgun proteomics analysis [16]. Several strategies, such as polyethylene glycol (PEG) fractionation [17], protamine sulphate precipitation [18], Rubisco IgY affinity [19] and polyethyleneimine-assisted Rubisco clean-up [16], have been used to deplete Rubisco from plant samples. Although protein precipitation is commonly used to remove high abundant proteins from plant samples, it can be difficult to re-solubilise proteins after precipitation leading to significant losses of protein yield [20].

During the depletion process, it is possible to lose other proteins due to non-specific binding to the depleted proteins and resin surfaces [21]. For example, comparative experiments between non-depleted serum and serum depleted of the six most abundant proteins have shown that while depletion significantly increased the number of proteins analysed and identified, some of the proteins found in the non-depleted serum were not found in the depleted serum [22, 23]. Therefore, when optimising protein depletion steps, the depleted protein resin bound fraction should be analysed to confirm that target proteins are not inadvertently lost.

1.2.4 Protein Enrichment

Enrichment strategies aim to enhance the concentration of particular classes of proteins (protein families or modified proteins, e.g. phosphoproteins) concomitantly reducing sample complexity for subsequent analysis and identification. This process is vital when the target proteins represent a minor component or have been post-translationally modified. Enrichment can be achieved by subcellular fractionation wherein centrifugation is the most straightforward protocol employed. Proteins located in the subcellular components, such as the membrane, mitochondria or nucleus, can be isolated to individual components based on their weight, size and shape through repeated centrifugation and velocity and equilibrium sedimentation [24, 25]. Centrifugation is also applicable to fractionation of protein complexes. Solvent- and/or chemical-based protein precipitation can be another choice to enrich low-abundance plant seed proteins [26].

Targeted strategies for enrichment include the use of IP, or through PTM-specific affinity binding, such as phosphoprotein enrichment [27]. These techniques are limited by the availability of suitable antibodies. During the binding process, proteins or peptides bind to the antibody or receptor surface while unrelated molecules remain free in the sample. Washing steps aim to eliminate non-specific binding, before the bound proteins are eluted from the receptors.

Electrophoretic methods can be used to separate proteins based on their size, shape and charge. Some of the most common electrophoretic pre-fractionation methods include electrokinetic methodologies which rely on isoelectric focusing (IEF) steps [28]. Frequently chromatography-based techniques are used in combination with subcellular fractionation approach to maximise the protein yield. For instance, subcellular fractionation technique combined with "loss-less" high-pH reversed-phase fractionation was used to isolate and extract low-abundant regulatory proteins from the human heart, mouse liver and brain [4, 29, 30].

To enrich post-translationally modified proteins from cell or tissue extracts, metal-ion affinity chromatography, titanium dioxide chromatography, superparamagnetic nanoparticles, affinity tags and high-affinity antibodies can be used [27, 31–33].

1.2.5   Normalisation

Cellular proteins span a wide dynamic range in biological samples. The most abundant components often dominate analyses hiding the minor components. Decreasing the concentration range to match the dynamic range of the analytical platform assists with the identification of low abundance proteins. Use of a combinatorial hexapeptide-bead library, for example, commercially available as ProteoMiner by Bio-Rad [34], can minimise the protein concentration dynamic range in serum and plasma samples [35]. This technique can reduce the amount of high-abundance proteins without immunodepletion while preventing the loss of low-quantity proteins bound to high-abundance proteins. The approach has also been extended to enrich for glycopeptides and phosphopeptides, which enables the study of post-translational modifications [36]. Use of this method for proteomics applications has been reviewed [35].

1.2.6   Protein Clean-Up

Protein isolation and extraction, enrichment or depletion steps introduce chemicals, such as detergents, salts, buffers and chaotropic agents, to improve protein solubility. The presence of these chemicals and biomolecules that coexist in the sample, e.g. lipids, can interfere with subsequent analysis, and their removal is not a straightforward process. Numerous methods have been described to remove detergents, chemicals and non-protein molecules from biological and clinical samples. For example, solid-phase extraction (SPE, e.g. StageTips and ZipTips), protein precipitation (e.g. trichloroacetic acid (TCA)/acetone), filters (ultrafiltration), chromatography-based techniques (high-pressure liquid chromatography (HPLC), fast protein

liquid chromatography (FPLC), size-exclusion, membrane, ion-exchange chromatography) and affinity-based techniques are frequently applied in labs [37–42].

In SPE-based techniques, compounds are separated based on their chemical and
physical properties, which determines their distribution between a mobile liquid
phase and a solid stationary phase. After protein (or peptide) binding, the remaining
compounds are washed away and the bound proteins are subsequently eluted from
the solid phase by altering the mobile phase. For example, StageTips contain very
small disks made of beads that are available in a variety of chemistries, reversed-
phase, cation-exchange or anion-exchange surfaces, embedded in a Teflon mesh,
which bind molecules based on their chemical nature [40]. Although SPE-based
approaches have been used in proteomics workflows, they are sample limited
(dependent on surface area/volume), are typically only capable of delivering crude
fractions from complex matrices, and may not give complete recovery required for
high-sensitivity experiments [43].

Precipitation buffers are used to remove interfering molecules from the proteomic samples [44]. TCA/acetone maintains the solubility of non-protein molecules while precipitating most proteins. Although protein precipitation is a rapid,
simple and inexpensive method, the inherent drawback for this method is the fact
that proteins may not fully resolubilise after TCA/acetone precipitation. Additionally,
this precipitation solution is unable to extract the maximum number of proteins
from recalcitrant plant tissues [26, 44].

Chromatography-based techniques are often included in the proteomic workflow
in order to enrich for target proteins and reduce sample complexity. HPLC can be
used online to separate the proteins/peptides and thereby increase the depth of
coverage. When employed offline, HPLC can purify or enrich proteins/peptides of
interest, typically using gradient elution and offline fraction collection. HPLC can
be tailored to particular protein classes, e.g. polar or non-polar protein separation.
Additionally ion-exchange chromatography is a complementary, but orthogonal,
technique capable of separating proteins on their charge state via ion-ion interactions
between the protein analyte and the stationary phase [45]. Additional to these
techniques, size-exclusion chromatography (SEC) can be used to separate proteins
based on their molecular size. SEC is a commonly used technique, because of the
diversity of the molecular weights of proteins in biological tissues and extracts. SEC
has been employed for many roles, including buffer exchange (desalting) or removal
of non-protein contaminants (DNA, viruses) [10].

Affinity chromatography, also known as affinity purification, is often used to characterise protein complexes and large protein interaction networks [46]. This technique
depends on the utilisation of sample-specific antigen-antibody recognition and binding.
The antibody is chemically immobilised or 'coupled' to a solid support so that when a
complex mixture is passed over the column, those molecules having the specific binding affinity to the antibody are captured. Other sample components are washed away
before the bound targets are eluted from the support, resulting in their enrichment from
the original sample. Affinity-tagged-based recombinant protein purification can be
used where proteins of interest are simply expressed in-frame with an epitope tag
(at either the N- or C-terminus), which is then used as an affinity handle to purify
the tagged protein (the bait) along with its interacting partners (the prey) [46].

1.3   Analytical Tools for Proteomics

A range of tools and methods are available to obtain information about proteins present in biological samples (Figs. 1.1 and 1.2). These diverse methods yield different types of information, and the choice of method depends on experimental requirements and/or the analytical goal. For example, the gel electrophoresis (GE) and WB techniques are applied to obtain macroscopic information about a protein, such as molecular weight, potential for dimerisation and oligomer formation, isoforms, relative abundance and protein conformation. MS offers great advantages in protein identification and quantitation (both relative and absolute), although antibody techniques such as ELISA are also capable of absolute quantitation. NMR spectroscopy and X-ray crystallography are most commonly employed for protein structural analysis. In this section, the commonly used analytical techniques will be discussed with reference to sample preparation methods that are used to maximise the value of each approach. The approaches include gel electrophoresis (1D; 2D; and difference in-gel electrophoresis (DIGE)), antibody-based techniques for targeting specific proteins or classes (WB, ELISA, IHC and antibody columns), MS-based high-throughput proteomics (top-down, middle-down and bottom-up) and the common strategies for data analysis and structural proteomics (NMR and X-ray crystallography).

1.3.1   Gel Electrophoresis

Gel electrophoresis (GE) separates protein or peptide mixtures on the basis of electric field strength, the net charge on the molecule, size and shape of the molecule, ionic strength and properties of the gel matrix (e.g. viscosity, pore size) [47]. Polyacrylamide and agarose are two background matrices commonly used in electrophoresis, where matrices serve as porous media and behave like a molecular sieve. Generally, GE can be performed within one dimension (SDS-polyacrylamide gel electrophoresis (PAGE) and native PAGE) and two dimensions (2D-PAGE and DIGE). After proteomic separation by GE, the proteins can be visualised by using various staining methods such as Coomassie dye-, silver-, zinc-, fluorescent- and functional group-specific stains [48].

One-dimensional (1D) electrophoresis separates proteins by mass, commonly used in the lab for comparative analysis of multiple samples. SDS-PAGE can be run in two modes: denaturing and reducing the protein with a discontinuous buffer system that separates the proteins based on their mass and non-denaturing PAGE, also known as native PAGE, which separates proteins according to their mass/ charge ratio. In the first method, protein samples are typically heated in the presence of SDS and dithiothreitol to denature the protein and the ionic detergent binds tightly to the protein to yield uniform negatively charged moieties. When a current is applied, all SDS-bound proteins within a sample migrate towards the positively charged electrode based on their mass. Proteins with known mass, known as markers, are run as a reference alongside samples in order to obtain the measurement of the size of the experimental protein. In native PAGE, proteins are run in an alkaline buffer, and they are separated according to the net charge, size and shape of their native structure. Unlike denaturing SDS-PAGE, no denaturants are used in this method, and thus subunit interactions within a multimeric protein are generally retained and information can be gained about the quaternary structure. Although this technique is used to prepare purified and active proteins, proteins with the same molecular weight cannot be separated and it is unsuitable for molecular weight determination.

Two-dimensional PAGE (2D-PAGE) is used for separation of complex protein mixtures by the independent parameters of isoelectric point and molecular weight [49]. The first dimension separates proteins according to their native isoelectric point using isoelectric focusing in a pH gradient. The second dimension separates the proteins based on their mass obtained from SDS-PAGE. 2D-PAGE provides the highest resolution for protein analysis and is an important technique in proteomic research, where the resolution of thousands of proteins on a single gel is sometimes necessary.

DIGE is a technique used to measure differential expression of proteins between two samples and a 1:1 mixture of the two samples on a single gel. This method is semi-quantitative and has three steps: the samples (individual or mixed) are each tagged with a fluorescent dye, run on a 2D-PAGE gel. The fluorescent images generated from scans of each dye of the same gel are superimposed over each other. In this way, DIGE allows relative quantitation of proteins by comparing each test sample to the control (mixture). The method can be extended to run additional samples all compared back to a universal standard as the control. Once protein spots of interest are identified, duplicate gels are run on a preparative scale and the spots of interest are matched, excised and subjected to in-gel digestion and LC-MS analysis for protein identification.

1.3.2 Antibody-Based Techniques

Antibody-based techniques commonly used include WB, ELISA and IHC. These techniques are highly dependent on the availability and quality of the antibody reagents and appropriate protein standards. Typically recombinant proteins are used as positive controls and standards. The antibody reagents may be monoclonal or polyclonal antibodies. Furthermore, good secondary antibody and amplification reagents to improve the detection of the antigen-bound primary antibody are critical. Generally secondary antibodies are anti-species immunoglobulin antibodies, for example, a rabbit anti-mouse immunoglobulin antibody. The secondary antibodies are commonly conjugated to enzymes such as horseradish peroxidase or alkaline phosphatase, which can be used with chromogenic, fluorogenic and chemiluminescent substrates. Biotinylated secondary antibody used in conjunction with enzyme labelled streptavidin can give excellent amplification of binding signal. All antibody-based techniques can suffer high backgrounds due to non-specific binding of any of the reagents. Frequently blocking buffers incorporating such reagents as non-fat dry milk, normal serum, BSA (bovine serum albumin), gelatin and one or more gentle surfactants are employed to reduce non-specific binding.

1.3.3   Western Blotting (WB)

WB is used to visualise a single protein or group of related proteins by using the ability of an enzyme- or fluorescence-conjugated primary antibody to bind to its specific antigen. Commonly labelled secondary antibodies are used to enhance the binding signal. The WB can be a rapid method to ascertain the presence of target proteins in complex protein mixtures and samples without investing much time in optimisation. Additionally it can inform if the target protein is well preserved or degraded in stored samples. The WB technique is inherently qualitative, but researchers often use this method semi-quantitatively to assess changes of apparent expression of the protein(s) of interest over time or as result of a perturbation. This method has three steps: gel electrophoresis (as described above) to separate the proteins according to size and/or charge; transfer to a membrane; and the antigen-antibody interaction and its detection with deposition of the chromogenic product at the site of antibody binding.

Samples for gel or WB analysis are prepared from a range of biological tissues and/or cells by the range of techniques described above (physical, chemical) with few limitations as the first stage of analysis (SDS-PAGE) is highly tolerant to chemicals and interfering components. Protease inhibitors are often added to the sample to prevent protein degradation during sample preparation. Prior to loading on the gel, protein concentrations of the samples are measured by one of many available assays (Bradford; Lowry; or bicinchoninic acid (BCA)) to ensure equivalent loadings of the amount of total protein, which is important if comparing samples. The denaturation of the sample proteins inherent in the SDS-PAGE methodology can efficiently expose linear epitopes making them available for binding to the primary antibody. Once the proteins are separated in the gel, the proteins are transferred to nitrocellulose or polyvinylidene fluoride (PVDF) membranes. Blocking reagents are used to minimise non-specific binding by the antibodies. The antibody binds to the antigen after a short incubation. In direct detection, an enzyme or biotin is linked directly to the primary antibody whereas indirect detection involves the use of a labelled secondary antibody.

1.3.4 ELISA

ELISA is a sensitive method used to detect the presence or quantify the abundance of an antigen. ELISAs are mostly conducted in 96- or 384-well microtiter plates. There are three main constructs of the ELISA: indirect, sandwich and competitive. For the indirect ELISA, the protein of interest, whether purified, enriched or in a crude preparation, is bound through adsorption, directly to the surface of the well, and subsequently detected by a combination of primary and labelled secondary antibodies. This simple approach can be a quick means of ascertaining the presence of a protein in a complex mixture or follow its enrichment through various separation protocols. The major limitation is low sensitivity; thus the target protein needs to be at relatively high abundance, for example, in the range of 0.2–5 µg/mL.

In the sandwich ELISA, a primary antibody is immobilised on the solid surface of a microtiter well. This antibody is often referred to as the 'capture' antibody. After a blocking step to eliminate any protein unadsorbed sites, the sample is added (usually diluted in detergent-containing buffers), and the target protein is allowed to bind to the primary antibody. After washing the unbound material, a second primary antibody is added and allowed to bind the bound antigen. This second primary antibody, which can be referred to as the 'detection' antibody, may be labelled and thus detected directly. Alternatively, the 'detection' antibody may be detected with labelled secondary antibodies if the two 'capture' and 'detection' antibodies are derived from different species (e.g. mouse and rabbit antibodies). Proteins of very low or unknown concentration in the sample can be detected because the 'capture' antibody essentially concentrates the low-abundance analyte. With the unbound material washed away, the binding of 'detection' antibody and subsequent amplification reagents are not hindered by interfering sample constituents.

Competitive ELISA can take many formats. The simplest is whereby the competitor antigen or protein is initially bound to the well, and after the blocking step the primary antibody is incubated with the sample with unknown amounts of the analyte or target protein. The well-bound antigen and the in-solution antigen derived from the sample compete for the primary antibody. Unbound material is washed away, and a labelled secondary antibody is added to the wells, which recognises the primary antibody in this case, captured by the well-adsorbed antigen. Unlike the other ELISAs and WB, decreased signal indicates high relative abundance of the target protein in the sample.

If high-quality preparations of the target protein are available to generate standard curves, both the sandwich and competitive ELISA formats can provide excellent quantitation.

1.3.5 Immunohistochemistry

IHC detects molecules (proteins) of interest within tissues using antibodies. Probes or enzymes conjugated to primary antibodies that bind the target molecule (or secondary antibodies that recognise the primary antibody) report the location of targets in tissue or cell samples for imaging. This method can help to visualise, identify and quantify the high-resolution distribution and localisation of specific proteins within cells and within their proper histological context [50]. Generally, this technique has two steps: firstly, sample preparation which broadly involves tissue collection and perfusion, fixation, embedding, sectioning and mounting, de-paraffinisation and epitope (antigen) retrieval, quenching/blocking endogenous target activity, blocking non-specific sites and sample labelling and, secondly, visualisation, interpretation and quantitation of the resultant protein expression [51].

1.3.6   Mass Spectrometry (MS)-Based Proteomics

MS-based proteomics represents a powerful tool for systems biology. It can provide a systematic, global, unbiased and quantitative assessment of proteins, including interactions, modifications and location, and can inform function [1, 6, 9]. Significant improvements in the sensitivity and the resolution of MS instrumentation and the associated advances in sample preparation and data analysis have dramatically advanced our understanding of the complex and dynamic nature of proteome. For example, MS-based proteomics can provide the quantitative measurement of the yeast proteome in 1 hour (of analytical time) [52]. New label-free approaches avoid some of the pitfalls and costs required to construct and analyse large quantities of tagged proteins or the generation of protein-specific antibodies [2]. Mass spectrometry can provide high-throughput analysis, offers high sensitivity and is capable of identifying and quantifying proteins spanning at least five orders of magnitude (dynamic range).

Based on the method of protein/proteome characterisation, MS-based proteomics approach can be divided into two main analytical streams: top-down and bottom-up approaches. Both methods have advantages and disadvantages for the identification of proteins and protein interacting partners, protein quantification and analysis of PTMs [53, 54]. Through top-down approach, intact proteins can be studied, which greatly simplifies sample preparation and reduces the mixture complexity as no proteolytic digestion is required [55]. Sample preparation for top-down proteomics also follows the classical proteomic sample preparation steps: protein isolation, extraction, denaturing, enrichment, gel electrophoresis and MS analysis. After extraction, chemical labelling can be included with soluble proteins for relative quantitation of samples by fluorescence (e.g. DIGE) or MS (e.g. isobaric tags for relative and absolute quantitation (iTRAQ) or tandem mass tags (TMT)) [56].

Top-down MS provides comprehensive information for the whole protein by detecting the range of PTMs (e.g. glycosylation, phosphorylation, methylation, acetylation) and sequence variants (e.g. mutations, polymorphisms, alternatively spliced isoforms) concurrently in one spectrum without a priori knowledge [57]. Although top-down MS can identify the intact proteins and their combination of modification events, this method is experimentally challenging for functional proteomics experiments due to the greater difficulty in analysing proteins in comparison with peptides. Proteins may be distributed across a number of isoforms [1]. Top-down methods have greater sample requirements, both in terms of protein quantity and purity. As a result, better separation methods are applied to reduce the sample complexity, obtain pure samples and increase the dynamic range of detection in top-down MS. Several offline and independent separation methods such as capillary HPLC, hydrophilic interaction liquid chromatography (HILIC) or combination of on- or offline HPLC with capillary isoelectric focusing, chromatofocusing or ion exchange can be used to separate and purify intact proteins for top-down MS analysis [58]. Furthermore, gel-eluted liquid fraction entrapment electrophoresis [59] and multicompartmental electrolyser called membrane-separated wells for isoelectric focusing and trapping [59] can also be included in the sample preparation to separate proteins for top-down-based MS analysis.

Independent of whether the target is protein or peptide, the analyte must be ionised and desorbed into the gas phase before detection and fragmentation in MS-based proteomics. Two types of ionisation techniques are used to generate the m/z in the gas phase: matrix-assisted laser desorption ionisation (MALDI) and electrospray ionisation (ESI). In MALDI samples are mixed with a matrix, often an organic acid, deposited on a target plate and ablated by a pulsed laser. The matrix absorbs energy transferring it to the analyte, leading to the ionisation of the analyte and its transfer into the gas phase. Generally, sample preparation for MALDI is straightforward and less complex than other MS-based methods, but there are still some points to be considered for successful sample preparation. Although MALDI is more tolerant to detergents and/or salts such as trifluoroacetic acid (TFA) or ammonium acetate which is often added to the LC for better separation, the presence of these chemicals can still interfere with crystallisation and/or ionisation. Additionally, the presence of higher amount of lipids and glycerol can interfere with the crystallisation process. The matrix selection can be empirically determined for each compound through a literature search or prior knowledge. Samples extracted for MALDI can be stored lyophilised or in solvent for future use. MALDI is most commonly coupled with time-of-flight (TOF) mass analysis. MALDI-TOF-MS has found great application downstream of one- or two-dimensional gel electrophoresis and more recently as mass spectrometry imaging (MSI) enabling detection and localisation of proteins in tissue samples [60]. This technique can also be a useful tool to explore molecular distribution and co-localisation of proteins in tissues or cells [61].

ESI is another soft ionisation technique used in MS for producing charged ions from proteins, peptides and other non-volatile macromolecules. In LC-MS-based proteomics, ESI provides high voltage to impart charge to the solvated analyte before exiting the electrode. The charged ions are desolvated and thus transferred to the gas phase before undergoing transmission and mass analysis. Once the molecule is in the gas phase, the m/z ratio of molecules is determined by their trajectories in a static or dynamic electric field. Taking advantages of high resolution and mass accuracy, top-down proteomics often uses ESI coupled to either Fourier transform ion cyclotron resonance (FT-ICR) [62] or Orbitrap mass analysers [63].

The bottom-up approach, also known as shotgun proteomics, relies upon proteolytic digestion of proteins into peptides with subsequent fragmentation by tandem mass spectrometry. Identification by in silico comparison of peptide fragment ion spectra to theoretical spectra generated from protein databases is accomplished using dedicated computational algorithms [1, 64]. In all of the bottom-up techniques, proteins were first extracted from the experimental sample and then digested into peptides with a suitable enzyme, most commonly trypsin. The peptides are then separated by HPLC and directed to the mass spectrometer for mass analysis.

Protein digestion with a dedicated enzyme is a critical element in bottom-up proteomics workflow. Trypsin is arguably the most commonly used enzyme in protein digestion due to its availability, high specificity and reproducibility. However, trypsin alone is unable to digest all extracted proteins, potentially missing sites of post-translational modifications or regions of proteins, leading to partial proteome coverage by bottom-up proteomics [65, 66]. To overcome these shortcomings, alternative proteases have been used such as chymotrypsin, LysC, LysN, AspN, GluC, LysargiNase and ArgC [65–67]. These alternative proteases are often used alone, combined and/or in a multi-protease strategy to increase sequence coverage, improve quantitative performance, specifically target post-translationally modified proteins, or to generate longer peptides for the 'middle-down' proteomics approach [66]. For example, the sole use of trypsin is insufficient to digest all the seed storage proteins (gluten family) from barley [68]. The study revealed that trypsin is the preferred enzyme for quantitation of avenin-like proteins, the B-, D- and γ-hordeins, while chymotrypsin is the enzyme of choice for the C-hordeins [68]. The use of multiple enzymes in proteomics sample preparation will increase protein sequence coverage, yielding greater depth of coverage, and uncover proteins that might otherwise elude detection.

Detergent-based [69] or detergent-free [70] in-solution digestion methods where proteins can be extracted either with Triton X-100, CHAPS or SDS or with strong chaotropic agents such as urea, thiourea and digestion of proteins under denaturing conditions are often used to purify and prepare proteins for bottom-up proteomics. Detergents are used regularly in proteomics to increase the solubility and stability and to disaggregate protein complexes [71], the detergent may interfere methods are useful to extract membrane-associated proteins [71], they can interfere with protein digestion, form adducts with peptides and proteins and ultimately increase the cleaning time for HPLC columns and sources and MS complexity, generating noisy baselines [72]. To overcome these pitfalls, there are LC-MS-compatible acid-labile surfactants such as RapiGest SF (Waters), Invitrosol™ (Thermo Fisher Scientific), ProteaseMAX™ (Promega) and PPS Silent® Surfactant (Agilent). These surfactant-based solutions can work in low pH and solubilise the proteins without interfering with the reversed-phase separations and MS.

In-solution digestion can be more readily automatable and minimises sample handling, but the proteome may be incompletely solubilised, and interfering substances may impede digestion. To overcome these shortcomings, filter-aided sample preparation (FASP) technique was introduced in combination with a molecular-weight cut-off filter [70]. This approach has a number of advantages over in-gel and in-solution digestion methods. For instance, FASP enables study of the whole proteome as a single fraction. By processing the proteins on a filter, harsh detergents or buffers may be used enabling greater protein solubilisation as the multiple buffer exchanges used in this approach can remove contaminants, maintain protein solubility, remove the incompatible reagents and thereby accommodate a range of digestion conditions.

MS-based bottom-up proteomics can be more suited to determine primary structure of protein, i.e. amino acid sequence. However, sequence variations which may have important biological implications may be missed when using this technique. Generally, MS-based bottom-up proteomics experiments can be divided into two broad categories: untargeted (qualitative/discovery) and targeted (quantitative) experiments. Discovery proteomics experiments are intended to identify as many proteins as possible across a broad dynamic range. This strategy often requires depletion of highly abundant proteins, enrichment of relevant components (e.g. subcellular compartments or protein complexes) and fractionation to decrease sample complexity (e.g. SDS-PAGE or chromatography). In contrast, targeted (quantitative) proteomics experiments are used to quantify a set of proteins with precision, sensitivity, specificity and high throughput. MS-based quantitative proteomics approaches can be further sub-divided into two types: relative and absolute quantitation experiments. Relative quantitation experiments aim to compare the spectral intensity or chromatographic peak area obtained from two or more analytical samples in order to determine the peptide abundance in one sample relative to another. These approaches can be label-free as is the case for multiple reaction monitoring (MRM) mass spectrometry or labelled experiments. Isotopic labels, such as stable isotope labelling with amino acids in cell culture (SILAC), iTRAQ or TMT can be used for relative quantitation experiments. Absolute quantitation requires the use of peptide standards, commonly employing heavy labelled isotopic peptides as an internal standard. Unlike relative quantitation, the abundance of the target peptide in the experimental sample is compared to that of the heavy peptide and back-calculated to the initial concentration of the standard using a predetermined standard curve to yield the absolute quantitation of the target peptide. In recent years, improvements in the analytical performance of MS instruments, including speed, sensitivity, mass accuracy and resolution, combined with new software algorithms have resulted in the ability to simultaneously identify and quantify thousands of proteins in very short analytical timeframes. Newly developed data-independent acquisition (DIA)-based methods such as SWATH [73] can simultaneously fragment a wide range of precursors that co-elute from the liquid chromatography column. The advantage of this method is that it is unbiased and minimises the stochastic nature of data sampling that occurs in data-dependent workflows.

A range of instrument configurations are available each suiting a different purpose. For instance, the triple quadrupole (QQQ) mass spectrometer is best suited to targeted quantitative proteomics because of their sensitivity and specificity [74]. In targeted proteomics, the proteins of interest are known; the aim is to generate robust, reproducible, sensitive and high-throughput acquisition of a subset of known peptides of interest [1].

Hybrid instruments, such as the quadrupole time of flight (Q-TOF), in which a quadrupole mass filter is coupled to a TOF mass analyser, or ion trap wherein ions are captured by the field allowing ion accumulation, impart greater mass accuracy and resolution. Frequently, linear ion traps are used in combination with an Orbitrap, in which ions circulate around a central, spindle-shaped electrode [75] which offers further increases in mass accuracy and sensitivity, enabling distinction of hundreds of thousands of different peptides from each other, which is often regarded as a precondition for their identification and quantitation [76].

Data analysis remains one of the major challenges in high-throughput MS-based experiments [1, 11, 74]. A multitude of software packages, vendor-specific and open-source, have been developed to analyse the discovery and targeted proteomics datasets. Open-source software, such as MaxQuant [77], Skyline [78] and OpenMS [79], can be used for bottom-up proteomics. While packages such as ProSight Lite [80] and MASH Suite Pro [81] can be used to analyse top-down proteomics data.

1.3.7 Structural Studies

Protein structure determination generates valuable information about protein folding, identification of ligand-binding sites, protein dynamics and their complexes with nucleic acids and RNA, which ultimately leads to a better understanding of their biological functions. NMR and X-ray crystallography remain the first choice for protein structural detection at high resolution, and both share similar sample preparation steps often requiring a large quantity of highly purified sample. The major bottleneck for structural studies lies in the need for high amounts of relatively pure samples, and hence sample preparation needs to generate recombinant proteins in a stable, soluble and suitably concentrated form for analysis [82]. The main difference between the two techniques is that NMR requires a sample in solution whereas X-ray crystallography requires crystalline samples.

Rapid screening and purification of expressed proteins for structural proteomics experiments can be accomplished using affinity and detection tags [83]. These tags are small in size so that they do not interfere with the target protein, and depending on the protein construct, the label can be removed before NMR and X-ray experiments [83, 84]. Chromatography-based techniques are frequently used to enrich or purify the protein samples before the NMR experiment.

NMR provides the information about angles, distances, coupling constants, chemical shifts, rate constants of a protein and suitable software that can generate the putative 3D structure [85]. NMR structure determination is limited currently by size constraints, lengthy data collection and analysis times (often months), and the method has limitations to detect the 3D structure for large monomeric and multimeric proteins due to ambiguous assignment of NMR signals [3]. The experimental cost can be high depending on the protein size and the complexity of structure elucidation.

Unlike NMR, X-ray crystallography uses X-ray diffraction (XRD) technologies to determine protein structural information at an atomic level. XRD experiments require a crystallisable protein but have no size limitations and provide the most precise atomic details [86]. However, this method is only applicable to the crystallised protein sample and is unable to examine the behaviour of the protein in the solution. Additionally, XRD uses one set of parameters at a time, and thus has the constraint of being able to examine only one confirmation at a time.

1.4 Summary

Despite the advancement of high-throughput technologies, data analysis and system-wide integration, significant challenges remain in being able to efficiently extract and purify proteins for downstream analysis. Protein extraction and purification plays a critical role in defining the results derived from any proteomic experiment. It is evident that no single sample preparation method is sufficient to extract all of the proteins from a given sample. Experiment-specific sample preparation method optimisation and analytical platform comparison is a prerequisite to establish the depth of coverage that can be obtained from complex biological matrices and define the limitations for proteomics is not a 'one size fits all' situtation, and consideration of sample type, amount and complexity and the overarching objectives of the study are needed. The subsequent chapters will deal with sample preparation from specific sample types providing a reference starting point that should be optimised in the hands and laboratories of the experimental scientist.

References

1. Aebersold R, Mann M (2016) Mass-spectrometric exploration of proteome structure and function. Nature 537:347
2. Larance M, Lamond AI (2015) Multidimensional proteomics for cell biology. Nat Rev Mol Cell Biol 16:269
3. Frueh DP, Goodrich AC, Mishra SH, Nichols SR (2013) NMR methods for structural studies of large monomeric and multimeric proteins. Curr Opin Struct Biol 23:734–739
4. Doll S, Dreßen M, Geyer PE, Itzhak DN, Braun C, Doppler SA et al (2017) Region and cell-type resolved quantitative proteomic map of the human heart. Nat Commun 8:1469
5. Selevsek N, Chang C-Y, Gillet LC, Navarro P, Bernhardt OM, Reiter L et al (2015) Reproducible and consistent quantification of the Saccharomyces cerevisiae proteome by SWATH-MS. Mol Cell Proteomics 14:739–749
6. Walther TC, Mann M (2010) Mass spectrometry–based proteomics in cell biology. J Cell Biol 190:491–500
7. Thein M, Sauer G, Paramasivam N, Grin I, Linke D (2010) Efficient subfractionation of gram-negative bacteria for proteomics studies. J Proteome Res 9:6135–6147
8. Abbott N (1988) Liquid-liquid extraction for protein separations. Chem Eng Prog 84:37

9. Jabbour R, Snyder A (2014) Mass spectrometry-based proteomics techniques for biological identification. In: Schaudies R (ed) Biological identification. Sawston, UK: Woodhead Publishing; p. 370–430
10. Tan SC, Yiap BC (2009) DNA, RNA, and protein extraction: the past and the present. J Biomed Biotechnol 2009
11. Baker MS, Ahn SB, Mohamedali A, Islam MT, Cantor D, Verhaert PD et al (2017) Accelerating the search for the missing proteins in the human proteome. Nat Commun 8:14271
12. Corthals GL, Wasinger VC, Hochstrasser DF, Sanchez JC (2000) The dynamic range of protein expression: a challenge for proteomic research. Electrophoresis 21:1104–1115
13. Chen YY, Lin SY, Yeh YY, Hsiao HH, Wu CY, Chen ST et al (2005) A modified protein precipitation procedure for efficient removal of albumin from serum. Electrophoresis 26:2117–2127
14. Liu G, Zhao Y, Angeles A, Hamuro LL, Arnold ME, Shen JX (2014) A novel and cost effective method of removing excess albumin from plasma/serum samples and its impacts on LC-MS/MS bioanalysis of therapeutic proteins. Anal Chem 86:8336–8343
15. Echan LA, Tang HY, Ali-Khan N, Lee K, Speicher DW (2005) Depletion of multiple high-abundance proteins improves protein profiling capacities of human serum and plasma. Proteomics 5:3292–3303
16. Zhang Y, Gao P, Xing Z, Jin S, Chen Z, Liu L et al (2013) Application of an improved proteomics method for abundant protein cleanup: molecular and genomic mechanisms study in plant defense. Mol Cell Proteomics 12:3431–3442
17. Xi J, Wang X, Li S, Zhou X, Yue L, Fan J et al (2006) Polyethylene glycol fractionation improved detection of low-abundant proteins by two-dimensional electrophoresis analysis of plant proteome. Phytochemistry 67:2341–2348
18. Kim YJ, Lee HM, Wang Y, Wu J, Kim SG, Kang KY et al (2013) Depletion of abundant plant R u B is CO protein using the protamine sulfate precipitation method. Proteomics 13:2176–2179
19. Cellar NA, Kuppannan K, Langhorst ML, Ni W, Xu P, Young SA (2008) Cross species applicability of abundant protein depletion columns for ribulose-1, 5-bisphosphate carboxylase/oxygenase. J Chromatogr B 861:29–39
20. Isaacson T, Damasceno CM, Saravanan RS, He Y, Catalá C, Saladié M et al (2006) Sample extraction techniques for enhanced proteomic analysis of plant tissues. Nat Protoc 1:769
21. Zolotarjova N, Martosella J, Nicol G, Bailey J, Boyes BE, Barrett WC (2005) Differences among techniques for high-abundant protein depletion. Proteomics 5:3304–3313
22. Yocum AK, Yu K, Oe T, Blair IA (2005) Effect of immunoaffinity depletion of human serum during proteomic investigations. J Proteome Res 4:1722–1731
23. Moritz RL, Ji H, Schütz F, Connolly LM, Kapp EA, Speed TP et al (2004) A proteome strategy for fractionating proteins and peptides using continuous free-flow electrophoresis coupled off-line to reversed-phase high-performance liquid chromatography. Anal Chem 76:4811–4824
24. Schubert OT, Röst HL, Collins BC, Rosenberger G, Aebersold R (2017) Quantitative proteomics: challenges and opportunities in basic and applied research. Nat Protoc 12:1289
25. Alberts B, Johnson A, Lewis J, Raff M, Roberts K, Walter P (2002) Fractionation of cells. Garland Science, New York
26. Smaczniak C, Li N, Boeren S, America T, Van Dongen W, Goerdayal SS et al (2012) Proteomics-based identification of low-abundance signaling and regulatory protein complexes in native plant tissues. Nat Protoc 7:2144
27. Oda Y, Nagasu T, Chait BT (2001) Enrichment analysis of phosphorylated proteins as a tool for probing the phosphoproteome. Nat Biotechnol 19:379
28. Righetti PG, Castagna A, Antonioli P, Boschetti E (2005) Prefractionation techniques in proteome analysis: the mining tools of the third millennium. Electrophoresis 26:297–319
29. Azimifar SB, Nagaraj N, Cox J, Mann M (2014) Cell-type-resolved quantitative proteomics of murine liver. Cell Metab 20:1076–1087
30. Sharma K, Schmitt S, Bergner CG, Tyanova S, Kannaiyan N, Manrique-Hoyos N et al (2015) Cell type–and brain region–resolved mouse brain proteome. Nat Neurosci 18:1819

31. Hwang L, Ayaz-Guner S, Gregorich ZR, Cai W, Valeja SG, Jin S et al (2015) Specific enrichment of phosphoproteins using functionalized multivalent nanoparticles. J Am Chem Soc 137:2432–2435
32. Aryal UK, Ross AR, Krochko JE (2015) Enrichment and analysis of intact phosphoproteins in Arabidopsis seedlings. PLoS One 10:e0130763
33. Larsen MR, Thingholm TE, Jensen ON, Roepstorff P, Jørgensen TJ (2005) Highly selective enrichment of phosphorylated peptides from peptide mixtures using titanium dioxide microcolumns. Mol Cell Proteomics 4:873–886
34. Thulasiraman V, Lin S, Gheorghiu L, Lathrop J, Lomas L, Hammond D et al (2005) Reduction of the concentration difference of proteins in biological liquids using a library of combinatorial ligands. Electrophoresis 26:3561–3571
35. Righetti PG, Boschetti E (2008) The ProteoMiner and the FortyNiners: searching for gold nuggets in the proteomic arena. Mass Spectrom Rev 27:596–608
36. Bandhakavi S, Van Riper SK, Tawfik PN, Stone MD, Haddad T, Rhodus NL et al (2011) Hexapeptide libraries for enhanced protein PTM identification and relative abundance profiling in whole human saliva. J Proteome Res 10:1052–1061
37. Hopper JT, Yu YT-C, Li D, Raymond A, Bostock M, Liko I et al (2013) Detergent-free mass spectrometry of membrane protein complexes. Nat Methods 10:1206
38. Lee SC, Knowles TJ, Postis VL, Jamshad M, Parslow RA, Lin Y-p et al (2016) A method for detergent-free isolation of membrane proteins in their local lipid environment. Nat Protoc 11:1149
39. Tubaon RM, Haddad PR, Quirino JP (2017) Sample clean-up strategies for ESI mass spectrometry applications in bottom-up proteomics: trends from 2012 to 2016. Proteomics 17:1700011
40. Rappsilber J, Mann M, Ishihama Y (2007) Protocol for micro-purification, enrichment, pre-fractionation and storage of peptides for proteomics using StageTips. Nat Protoc 2:1896
41. Bladergroen MR, van der Burgt YE (2015) Solid-phase extraction strategies to surmount body fluid sample complexity in high-throughput mass spectrometry-based proteomics. J Anal Methods Chem 2015:250131
42. Ghosh R (2002) Protein separation using membrane chromatography: opportunities and challenges. J Chromatogr A 952:13–27
43. Yu Y, Smith M, Pieper R (2014) A spinnable and automatable StageTip for high throughput peptide desalting and proteomics. Protoc Exchange. https://doi.org/10.1038/protex.2014.033
44. Wu X, Xiong E, Wang W, Scali M, Cresti M (2014) Universal sample preparation method integrating trichloroacetic acid/acetone precipitation with phenol extraction for crop proteomic analysis. Nat Protoc 9:362–374
45. Cummins PM, Rochfort KD, O'Connor BF (2017) Ion-exchange chromatography: basic principles and application. Methods Mol Biol 1485:209–223
46. Gingras A-C, Gstaiger M, Raught B, Aebersold R (2007) Analysis of protein complexes using mass spectrometry. Nat Rev Mol Cell Biol 8:645
47. Gibson F, Anderson L, Babnigg G, Baker M, Berth M, Binz P-A et al (2008) Guidelines for reporting the use of gel electrophoresis in proteomics. Nat Biotechnol 26:863
48. Steinberg TH (2009) Protein gel staining methods: an introduction and overview. In: Burgess MPD RR (ed) Methods in enzymology, vol 463. Elsevier, pp 541–563
49. Wilkins MR, Pasquali C, Appel RD, Ou K, Golaz O, Sanchez J-C et al (1996) From proteins to proteomes: large scale protein identification by two-dimensional electrophoresis and amino acid analysis. Nat Biotechnol 14:61
50. Xing Y, Chaudry Q, Shen C, Kong KY, Zhau HE, Chung LW et al (2007) Bioconjugated quantum dots for multiplexed and quantitative immunohistochemistry. Nat Protoc 2:1152
51. De Matos LL, Trufelli DC, De Matos MGL, da Silva Pinhal MA (2010) Immunohistochemistry as an important tool in biomarkers detection and clinical practice. Biomark Insights 5:9–20
52. Hebert AS, Richards AL, Bailey DJ, Ulbrich A, Coughlin EE, Westphall MS et al (2013) The one hour yeast proteome. Mol Cell Proteomics 13:339–347

53. Tran JC, Zamdborg L, Ahlf DR, Lee JE, Catherman AD, Durbin KR et al (2011) Mapping intact protein isoforms in discovery mode using top-down proteomics. Nature 480:254
54. Catherman AD, Skinner OS, Kelleher NL (2014) Top down proteomics: facts and perspectives. Biochem Biophys Res Commun 445:683–693
55. Moradian A, Kalli A, Sweredoski MJ, Hess S (2014) The top-down, middle-down, and bottom-up mass spectrometry approaches for characterization of histone variants and their post-translational modifications. Proteomics 14:489–497
56. Padula MP, Berry IJ, Raymond B, Santos J, Djordjevic SP (2017) A comprehensive guide for performing sample preparation and top-down protein analysis. Proteomes 5:11
57. Siuti N, Kelleher NL (2007) Decoding protein modifications using top-down mass spectrometry. Nat Methods 4:817–821
58. Cui W, Rohrs HW, Gross ML (2011) Top-down mass spectrometry: recent developments, applications and perspectives. Analyst 136:3854–3864
59. Meng F, Cargile BJ, Patrie SM, Johnson JR, McLoughlin SM, Kelleher NL (2002) Processing complex mixtures of intact proteins for direct analysis by mass spectrometry. Anal Chem 74:2923–2929
60. Susnea I, Bernevic B, Wicke M, Ma L, Liu S, Schellander K et al (2012) Application of MALDI-TOF-mass spectrometry to proteome analysis using stain-free gel electrophoresis. In: Cai ZLS (ed) Applications of MALDI-TOF spectroscopy. Topics in current chemistry, vol 331. Springer, Berlin, Heidelberg, pp 37–54
61. Cornett DS, Reyzer ML, Chaurand P, Caprioli RM (2007) MALDI imaging mass spectrometry: molecular snapshots of biochemical systems. Nat Methods 4:828–833
62. Li H, Nguyen HH, Loo RRO, Campuzano ID, Loo JA (2018) An integrated native mass spectrometry and top-down proteomics method that connects sequence to structure and function of macromolecular complexes. Nat Chem 10:139
63. Bogdanov B, Smith RD (2005) Proteomics by FTICR mass spectrometry: top down and bottom up. Mass Spectrom Rev 24:168–200
64. De Godoy LM, Olsen JV, Cox J, Nielsen ML, Hubner NC, Fröhlich F et al (2008) Comprehensive mass-spectrometry-based proteome quantification of haploid versus diploid yeast. Nature 455:1251
65. Giansanti P, Tsiatsiani L, Low TY, Heck AJ (2016) Six alternative proteases for mass spectrometry–based proteomics beyond trypsin. Nat Protoc 11:993
66. Tsiatsiani L, Heck AJ (2015) Proteomics beyond trypsin. FEBS J 282:2612–2626
67. Huesgen PF, Lange PF, Rogers LD, Solis N, Eckhard U, Kleifeld O et al (2014) LysargiNase mirrors trypsin for protein C-terminal and methylation-site identification. Nat Methods 12:55–58
68. Colgrave ML, Byrne K, Howitt CA (2017) Food for thought: selecting the right enzyme for the digestion of gluten. Food Chem 234:389–397
69. Nagaraj N, Lu A, Mann M, Wiśniewski JR (2008) Detergent-based but gel-free method allows identification of several hundred membrane proteins in single LC-MS runs. J Proteome Res 7:5028–5032
70. Wiśniewski JR, Zougman A, Nagaraj N, Mann M (2009) Universal sample preparation method for proteome analysis. Nat Methods 6:359
71. Laganowsky A, Reading E, Hopper JT, Robinson CV (2013) Mass spectrometry of intact membrane protein complexes. Nat Protoc 8:639
72. Cañas B, Piñeiro C, Calvo E, López-Ferrer D, Gallardo JM (2007) Trends in sample preparation for classical and second generation proteomics. J Chromatogr A 1153:235–258
73. Gillet LC, Navarro P, Tate S, Röst H, Selevsek N, Reiter L et al (2012) Targeted data extraction of the MS/MS spectra generated by data-independent acquisition: a new concept for consistent and accurate proteome analysis. Mol Cell Proteomics 11:O111. 016717
74. Gillet LC, Leitner A, Aebersold R (2016) Mass spectrometry applied to bottom-up proteomics: entering the high-throughput era for hypothesis testing. Annu Rev Anal Chem 9:449–472
75. Hager JW (2002) A new linear ion trap mass spectrometer. Rapid Commun Mass Spectrom 16:512–526

76. Hu Q, Noll RJ, Li H, Makarov A, Hardman M, Graham Cooks R (2005) The Orbitrap: a new mass spectrometer. J Mass Spectrom 40:430–443
77. Cox J, Mann M (2008) MaxQuant enables high peptide identification rates, individualized ppb-range mass accuracies and proteome-wide protein quantification. Nat Biotechnol 26:1367
78. MacLean B, Tomazela DM, Shulman N, Chambers M, Finney GL, Frewen B et al (2010) Skyline: an open source document editor for creating and analyzing targeted proteomics experiments. Bioinformatics 26:966–968
79. Sturm M, Bertsch A, Gröpl C, Hildebrandt A, Hussong R, Lange E et al (2008) OpenMS–an open-source software framework for mass spectrometry. BMC Bioinformatics 9:163
80. Fellers RT, Greer JB, Early BP, Yu X, LeDuc RD, Kelleher NL et al (2015) ProSight Lite: graphical software to analyze top-down mass spectrometry data. Proteomics 15:1235–1238
81. Cai W, Guner H, Gregorich ZR, Chen AJ, Ayaz-Guner S, Peng Y et al (2015) MASH suite pro: a comprehensive software tool for top-down proteomics. Mol Cell Proteomics 15:703–714
82. Yee AA, Savchenko A, Ignachenko A, Lukin J, Xu X, Skarina T et al (2005) NMR and X-ray crystallography, complementary tools in structural proteomics of small proteins. J Am Chem Soc 127:16512–16517
83. Elsliger M-A, Deacon AM, Godzik A, Lesley SA, Wooley J, Wüthrich K et al (2010) The JCSG high-throughput structural biology pipeline. Acta Crystallogr Sect F Struct Biol Cryst Commun 66:1137–1142
84. Gräslund S, Nordlund P, Weigelt J, Hallberg BM, Bray J, Gileadi O et al (2008) Protein production and purification. Nat Methods 5:135
85. Banci L, Bertini I, Luchinat C, Mori M (2010) NMR in structural proteomics and beyond. Prog Nucl Magn Reson Spectrosc 56:247
86. Krishnan V, Rupp B (2012) Macromolecular structure determination: comparison of x-ray crystallography and nmr spectroscopy. eLS 10:a0002716

Chapter 2
Sample Treatment for Saliva Proteomics

Francisco Amado, Maria João Calheiros-Lobo, Rita Ferreira, and Rui Vitorino

*"Saliva
doesn't have the drama of blood,
it doesn't have the integrity of sweat and
it doesn't have the emotional appeal of tears."*
—Irwin Mandel (professor emeritus, Columbia University)

2.1 Introduction

The analysis of saliva can offer an approach with a good cost-effectiveness for screening diseases in large populations, as well as use in children and in elderly, where other types of samples present further complications [20, 71]. Compared with other body fluids, such as blood, plasma, serum, or urine, saliva constitutes an alternative for diagnostic with clinical and toxicology purposes [95].

Saliva is normally easily obtained in analytically practical amounts, with simple, safe, painless, and noninvasive collection procedures in a much more patient friendly way when compared with blood sampling, making follow-up studies more feasible. Probably, the major limitation of saliva as diagnostic sample is the interindividual variability in terms of composition with different salivary flows and water content (e.g., individualized protein concentrations), viscosity, and differentiated contributions of cellular exudates/transudates making difficult the comparison

F. Amado (✉) · R. Ferreira
QOPNA/LAQV, Departamento de Química da Universidade de Aveiro, Aveiro, Portugal
e-mail: famado@ua.pt

M. J. Calheiros-Lobo
CESPU, IINFACTS, Departamento das Ciências Dentárias do IUCS, Gandra PRD, Portugal

R. Vitorino
iBiMED, Department of Medical Sciences, Portugal, University of Aveiro, Aveiro, Portugal

© Springer Nature Switzerland AG 2019
J.-L. Capelo-Martínez (ed.), *Emerging Sample Treatments in Proteomics*, Advances in Experimental Medicine and Biology 1073, https://doi.org/10.1007/978-3-030-12298-0_2

between patients [95]. Nevertheless, saliva composition reflects local and systemic pathophysiological conditions [82], which allow to perspective a rising use of oral saliva-based biomarker for point-of-care testing and future development of lab-on-a-chip-based technology [7, 89]. Salivary proteomics hold a special promise in disclosing new potential salivary biomarkers for oral and systemic diseases [20, 82, 92]. A good example that seems highly relevant for salivary testing is head and neck cancer [59, 129]. Other examples are periodontitis [50, 89, 133], dental caries [141, 142], or Sjörgen's syndrome [31].

For non-oral systemic diseases, a review of literature suggests that human salivary proteomics has been successfully employed in diagnostic of diseases such as acute myocardial infarction [89]; type 1 diabetes [18, 19, 61]; type 2 diabetes [61, 109]; breast [88, 126, 127], prostate [116], and ovarian cancers [22, 37]; viral infections [35]; hereditary diseases [10]; and autoimmune diseases[98], among others (for a review on this subject, see [26]).

The aim of this chapter is to highlight crucial procedures for successful salivary proteomics such as sample collection, handling and storage, and to give a glimpse into the factors that influence the variability of sample composition due to technical and subject issues.

2.2   Saliva Secretion and Composition

Healthy adults produce at rest between 0.5 and 1.5 liters of saliva per day (or approximately 0.5 mL/min) [23] that may easily increase more than double under stimulation [90]. Saliva is a hypotonic fluid composed mostly of water, electrolytes, and biomolecules such as proteins [89]. Saliva is vital in the maintenance of oral tissues health with mucosal and teeth protection proprieties including lubrication and hydration; pH buffering; protection from dental erosion/demineralization; protection against pathogenic microbiota, namely, by direct antimicrobial actions [135] or preventing the adhesion of microorganisms to oral tissues; and facilitating oral functions such as speaking, swallowing, food tasting, mastication, and initiation of digestion. Moreover, saliva composition reflects systemic health status [16].

Saliva is produced by salivary glands, composed of serous and mucous cells (acinar cells) and of different types of duct cells, contributing differently to saliva composition. Saliva results from the secretion of three pairs of major glands: the parotid (almost exclusively of serous cells; contributing around 60%), the submandibular (contains both types of cells, with the serous somewhat more numerous; 20%), and the sublingual (the majority of the acini are mucous cells; 5%) [63]. In addition to the major salivary glands, there are hundreds (500 to 1000) of minor glands (mucous cells) located in the lips, tongue, palate, and cheeks, which contribute in about 15% to saliva composition [121].

Secretion of saliva is an active and continuous process mainly under the influence of the sympathetic nervous system. Any autonomic nervous system disturbance will easily lead to derangement, frequently dominated by abnormal storage and acinar swelling [86]. Acinar cells synthesize large quantities of proteins, which are combined with imported water, salts, and various other components derived from plasma to produce saliva. Duct cells contribute to saliva final composition importing plasma components and producing some proteins such as growth factors, immunoglobulins, and kallikreins [3, 20, 140]. In addition to secreted saliva (>90%), there are other oral and systemic contributions that all together constitutes what is called whole saliva. Whole saliva is, in fact, a complex mixture which comprises several components, such as gingival crevicular fluid, oral mucosa exudate/transudate, oral mucosal cells, nasal and bronchial secretions, serum filtrate, wound blood products (directly to the oral cavity or via a transepithelial route) or oral diseases contributions, multiple origin exosomes, components of the complex oral microbiota (viruses, fungi, bacteria), or even food debris [3, 34, 140].

A wide variety of proteins have been found in this oral fluid. To date, more than 3000 different protein species have been identified in human saliva using proteomic approaches [5, 104, 115, 157]. The bulk (90%) of all saliva proteins comprises a polymorphic group of proteins synthesized by the salivary glands: mucins (large glycoproteins), proline-rich proteins (PRP), histatins, tyrosine-rich proteins, statherin, and anionic and cationic glycoproteins [3]. Many other proteins have been identified mostly originated from oral tissues and plasma. Identified protein species are very heterogeneous going from high molecular weight glycoproteins (mucins), to a high percentage of peptides and small proteins, many arriving from posttranslational proteolytic cleavage of the precursor forms. It should be noted that 20–30% of all identified protein species belong to the main salivary peptide classes, namely, statherin, PRPs, and histatins [5]. Regarding peptides and small proteins (MW < 16,000 Da) more than 2000 species were already identified using proteomics/peptidomics. Most of them are originated from cellular debris or plasma components, suggesting a high proteolytic activity inside the oral cavity [145].

The salivary composition depends on the contribution of each salivary gland. For example, the basic proline-rich proteins are secreted only by the parotid glands, while cystatin S-type is mainly excreted by the submandibular and sublingual glands; the acidic proline-rich proteins and statherin are secreted by all glands, although in different relative amounts [5]. The mucins, high molecular weight glycoproteins (such as MUC5B and MUC7), originate mainly from mucous cells [86] and consequently make saliva from sublingual, submandibular, and minor glands viscous and difficult to technically process [63]. Mucins present a high degree of glycosylation and hydration potential, able to prevent dehydration and provide the necessary lubrication of oral cavity [34, 143]. Although bacteria are commonly referred as part of saliva, bacterial proteins identification in saliva is limited, being only possible when multidimensional approaches are used [146, 147].

2.3   Saliva Collection

Saliva collection approaches and the most used commercial devices will be presented and discussed in this section (Fig. 2.1). Focus will be given to the influence of the chosen methodology on saliva flow and how it interferes with the contribution

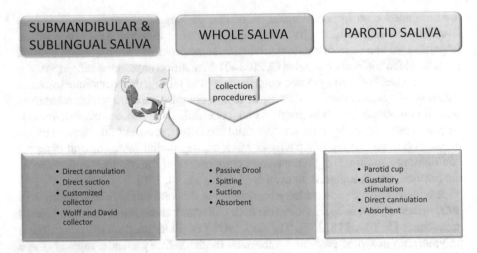

Fig. 2.1 Main saliva collection procedures for submandibular/sublingual glands; whole saliva; and parotid saliva

of each of the different salivary glands for the sample composition, being aware that nowadays saliva collection is not standardized.

Depending on the methodology used for saliva collection, different types of saliva might be considered [9]:

(i) Whole saliva also called mixed saliva or oral fluid (unstimulated saliva): correspond to the sum of all possible contributions to the collected fluid, namely, the gingival crevicular fluid, oral microbiota and their metabolic products, mucosal cell debris, nasal discharge, gingival crevicular fluid, and food debris.

(ii) Parotid saliva: fluid secreted by the parotid glands and obtained directly from the parotid duct orifice.

(iii) Submandibular saliva: fluid secreted by the submandibular glands and obtained directly from the submandibular duct orifice.

(iv) Sublingual saliva: fluid secreted by the sublingual glands and obtained directly from the sublingual duct orifice.

(v) Submandibular/sublingual saliva: fluid secreted by the submandibular and sublingual glands and obtained directly from the floor of the mouth in the vicinity of the submandibular duct opening, when secretion from the parotid glands is prevented.

(vi) Minor salivary gland secretions: the fluids secreted by the minor salivary glands and obtained directly from the duct openings. The location of the glands should be stated (e.g., labial, palatine).

(vii) Stimulated saliva: all the above types of saliva may be collected with increased excretion rates after gustatory, masticatory, pharmacologic, or mechanical stimulation.

2.3.1   Saliva Collection Procedures

2.3.1.1   Whole Saliva Collection

Whole saliva (unstimulated) is mainly collected by four different methods: the passive drool method, the spitting method, the suction method, and the absorbent method.

The Draining (Passive Drool) Method

For collection from subjects in resting and awake waking, saliva is let to accumulate in the floor of the mouth with forward tilted head, and saliva is allowed to drop continuously off the lower lip into a collection tube (a funnel may help) [48, 93], restricting any oral movement [87]. In theory this is probably the best method since it avoids any kind of bias such as reflex stimulation or different contributions from salivary glands.

The Spitting Method

As a variant of the above methodology, after allowing the saliva to accumulate in the mouth the subject is asked to spit saliva into the collection tube [93, 151]. Subjects spit into the collection tube about once a minute [151]. In this method a higher bacterial contamination of the sample is expected [49].

The Suction Method

Another possibility is to use a small aspirator device and continuously withdraw saliva from the floor of the mouth [87, 93]. As expected and according to Michishige et al. [87], suction strongly stimulates saliva secretion.

The Absorbent Method

In the absorbent method, saliva is collected by a cotton roll, swab, or foam material placed in the mouth [27] that then is removed from the oral cavity, and the final sample obtained after centrifugation of the adsorbed material.

The use of absorbent devices may be advisable in large studies using many different people involved in collection to avoid operator errors. Moreover, their use is mandatory in the case of small children or individuals with motor or sensitive disabilities that have difficulty with the passive drool technique. In this approach the location of the absorbent in the mouth is crucial since there may be a differentiated contribution from each gland. Nevertheless, the results obtained from the saliva of the swab placed underneath the tongue should be similar to those from whole saliva collected by passive drool.

Presently, there are commercial devices available for the absorbent method [120, 121] (that will be discussed in more detail in Sect. 2.3.1.3) used with small technical differences. After removing the adsorbent from the oral cavity, saliva may be transferred into a collection vial by centrifugation (e.g., Sarstedt, British Company Malvern Medical Developments, or Salimetrics general saliva collection devices) or by mechanical pressure trough a syringe plunger (e.g., Oasis Diagnostics saliva collection kits) [85].

Note: It is necessary to remember that due to the potential for suffocation saliva collection from infants requires special consideration.

2.3.1.2 Stimulated Saliva

All types of saliva may be collected with increased excretion rates after gustatory, masticatory, pharmacologic, or mechanical stimulation. The two most used procedures are gustatory stimulation, with acidic solutions (e.g., citric acid), and mechanical stimulation by chewing. Stimulation with citric acid is the most used methodology with variable solution concentrations that may go from 0.25% [124] to 3% [1] or even citric acid powder in a swab [151], with repeated applications over the tongue every 30 s for diluted acid concentrations [1] or every 2 min for higher ones [151]. According to Stokes and Davies [124], in the case of whole saliva, both acidic solutions and mechanical action allow the collection of similar volumes of saliva; however, mechanical action stimulates slightly shear-thinning and relatively inelastic saliva, while acidic solutions stimulate secretion of saliva that is highly elastic and shear-thinning. The variation of collected saliva rheology occurs due to the different proportion of saliva secreted from each gland depending on the method of stimulation [124]. Other discrepancies due to different subject characteristics such as oral buffer capacities, latency time, or blood perfusion conditions, are expected. Giving the latency time to stabilize secretion when salivary glands are stimulated the initial fluid should be discarded (at least the first minute). Many salivary proteins are differentially secreted with acid stimulation, thus influencing the salivary proteome composition [156], a fact that should not be forgoten.

In practical terms there are two situations for which stimulation is worthwhile: when not enough saliva can be collected without their use (e.g., xerostomic patients) or in the case of saliva collection from parotid glands that present low saliva flow rates in rest conditions. It is possible to point some additional advantages for the collection of stimulated saliva, as it allows a better standardization of salivary flow in a heterogeneous group of subjects, is faster, and is more convenient for subjects [92].

Stimulants can be used but sparingly and in a consistent manner throughout the entire experiment since they may exacerbate interindividual variation and changes in saliva composition.

2.3.1.3   Commercial Devices for Saliva Collection

There are several trading companies that sell saliva diagnostic tools focused in test kits with specific applications like DNA collection, HIV, hormones, or abuse substances tests, such as OraSure™ technologies (e.g., OraSure Oral Specimen Collection Device, http://www.orasure.com/products-insurance/products-insurance.asp), StatSure Diagnostic Systems (e.g., Quantisal Oral Fluid Collection Device), Greiner Bio-One (Kremsmünster, Austria, www.gbo.com), Sarstedt (Germany, www.sarstedt.com), Salimetrics (State College, PA, https://www.salimetrics.com), or Oasis Diagnostics® Corporation (Vancouver WA USA, www.4saliva.com). This is clearly a growing market.

In the saliva collection field, devices that may be used in proteomics include Salivette™ introduced by the Sarstedt company (Germany, www.sarstedt.com) in 1987. Currently, Salivette™ is available with two kinds of swabs, a plain cotton swab and a cotton swab with citric acid to stimulate salivation, both coupled with a special conical polypropylene centrifuge tube that allows separating mucous and particles, obtaining a clear sample without further sample handling. To collect a sample, patients place the swab in the mouth and chewed for approximately 1 min (https://www.sarstedt.com/en/products/diagnostic/salivasputum/).

SalivaBio (Salimetrics, State College, PA, https://www.salimetrics.com) relies on polypropylene vials for the passive drool methodology. This device should be used with the Saliva Collection Aid (also from Salimetrics), a plastic funnel-type device, to sample up to 2 mL of saliva. Salimetrics also produces a swab collection system such as the Salimetrics Oral Swab (SOS) device. This SOS device uses an inert polymer material pad as collection medium. The sample is collected by placing the absorbent pad in the mouth of the pediatric patients from 1 to 5 min, after which the pad is placed into the conical tube provided. Other options are also available for children and infants, the SalivaBio Children's Swab (SCS) and the SalivaBio Infant's Swab (SIS), respectively, with smaller swabs.

The Malvern Medical Developments company (www.malmed.co.uk) developed the ORACOL™ and ORACOL PLUS™ Collection Kits that uses an absorbent foam material in a swab format to collect up to 1 mL of whole saliva into a centrifuge tube. The ORACOL™ swab is placed in the mouth and allowed to absorb saliva. The sample is winkled out by centrifugation from the swab using a tube provided in the kit. In the ORACOL PUS™ option, a microtube is incorporated within the device so that the saliva is centrifuged directly into the microtube provided.

Oasis Diagnostics™ Corporation (Vancouver, WA, USA, www.4saliva.com) manufactures a series of oral-based tools, which includes the Versi•SAL™ device for standardized whole saliva collection, the UltraSal-2™ for large volume oral specimen collection, the Super•SAL™ saliva collection device for universal saliva collection purposes, and the RNAPro•SAL™, a device for RNA and/or protein collection for genomic or proteomics applications. The "Super•SAL™ universal saliva collection kit" allows the collection of whole saliva using a highly absorbent cylindrical-shaped noncellulosic pad giving typically volumes higher than 1.0 mL in approximately 1 to 3 min and includes a sample volume adequacy indicator that turns from yellow to blue whenever a sufficient sample has been collected. Collected saliva is then separated by compressing the absorbent pad through a compression chamber (syringe plunger) into a standard 2 mL Eppendorf tube or a 1.5 mL microfuge tube. The trade company presents three variations of Super•SAL™: the Micro•SAL™, a Children Saliva Collection Kit adapted for the collection of saliva samples from younger children with a small soft pad, collecting up to a maximum of 500 µl of saliva; the Pedia•SAL™ Infant Salivary Collection Kit that integrates a perforated pacifier with the rest of the saliva sampling kit; and the Pure•SAL™ Oral Specimen Collection Kit which includes a filter in the compression tube that removes additional interferents and large molecules and according to the manufacturer is suitable for the isolation of cell-free DNA, cell-free RNA, exosomes, or proteins in a single step. Versi•SAL™ is a similar kit over a maximum sample volume of 1.4 mL, with a different shape pad, to collect saliva from under the tongue. A variation includes a modified compression tube that split sample into two sample tubes simultaneously, allowing to obtain two samples from the same patient. The RNAPro•SAL™ kit was developed as a device for standardized collection of saliva RNA and proteins providing two equivalent samples of saliva, through a splitting unit attached to the compression tube, for a total of 1.0 mL of saliva in 1–3 min, being also suitable for the isolation of exosomes and the use of cell-free DNA or cell-free RNA. Another option is Accu•SAL™ designed for saliva standardized collection, which incorporates graduated scale on the side of the transport tube. Lastly, the UltraSal-2™ saliva collection kit is used for the collection of up to 24 mL of whole saliva by the drool technique. The UltraSal-2™ kit includes two collection tubes of 12 mL each connected to a single mouth piece into which the user expectorates.

Since most of the devices are devoid from any preservative agents, samples must be centrifuged and rapidly preserved prior to analysis. For a review on this subject you may want to consult the work of Slowey [121].

2.3.1.4   Salivary Gland Saliva Collection

Saliva can be selectively collected from individual salivary glands using the aid of specially constructed collectors while blocking saliva drainage from the ducts of the other glands normally by a cotton gauze (not mandatory for parotid saliva collection).

The Collection of Parotid Saliva

Saliva collection from an isolated parotid gland saliva is the easiest of the individual glandular secretions to collect. It can be collected with the use of a parotid cup [151] (a plastic container stabilized on the mucosal surface by a negative pressure enabling pocket) placed faced to the oral mucosa, between the cheek and upper gum at the level of the second upper molar, where the parotid duct (Stenson's duct) opens into the oral cavity. The parotid cup is a device known as the Carlson–Crittenden collector, originally reported in 1910 [15, 60], popularized by Lashley in 1916 (also known as the Lashley cup) [76], consisting of two concentric chambers communicating with the exterior by means of two cannulae. The central chamber provides an exit for parotid saliva and vacuum is applied in the external compartment in order to maintain the device in place [85]. Samples are collected via suction onto ice using an induced stimulation (typically a sterile aqueous citric acid solution applied on the tongue by means of a cotton swab at periodic intervals). The application of suction cups mounted simultaneously on both parotid ducts is desirable to increase yield and shorten collection times [68, 121].

Since parotid glands present low flow rates in rest conditions, in addition to a bilateral collection, a simple gustatory stimulus such as a citric acid solution applied on the tongue surface, by means of a gauze pellet every 30 sec [86], is advisable. Depending on the study design, the first 0.1 mL of collected saliva should be discarded to ensure that fresh parotid saliva is obtained [151] and also to compensate latency time.

Alternatively, enriched parotid saliva may well be collected using an absorbent device or using direct cannulation of the parotid duct [86] and a thin tube is placed directly at the outlet of the main parotid excretory duct (Stensen's duct; in this case help from a dental health professional is mandatory). This method may induce discomfort and requires a skilled operator. In some cases, application of a local anesthetic is required [121].

Collection of Human Submandibular (SM) and Sublingual (SL) Saliva

Several processes exist to collect saliva from the SM and SL glands simultaneously. It should be noted that separate collection of SM/SL is difficult because of the close anatomical relation between the orifices of the two glands and the common presence of communicating ducts between the submandibular and sublingual main ducts [68]. Given that, it is difficult to collect the fluids separately, so often the option is for a joint collection and, in this case, the concern will be to isolate the saliva of the SM and SL glands from other contributions in particular of the parotid gland by blocking the parotid ducts with a cotton roll. Since the problems of separate collection are similar for the two types of glands these will be treated together. It is possible to collect unstimulated secretions but as for parotid saliva collection, some sort of stimulus is often applied (usually a citric acid solution applied directly to the tongue). SM/SL collecting methods can be divided into Wolff and Davis collector method, direct cannulation methods, direct suction methods, and customized collectors method.

The Wolff and Davis Collector Method

Most probably the more reliable and simple SM/SL saliva collection system is the Wolff and Davis device [152], consisting of four parts: collecting tubing, a buffering chamber, a storing tube, and a suction device. A high yield (90%) of relatively pure SM/SL fluids is obtainable with minimal contamination. Using slightly modified procedures, the system may be optimized to collect either specimen type [153].

In practice, for submandibular saliva collection, each parotid duct is typically blocked using cotton gauze, the floor of the mouth is then dried, and the openings to the sublingual glands on both sides of the mouth are also blocked. The subject should raise their tongue slightly to elevate the opening of the SM gland and collection of SM saliva performed using a sterilized Wolff device [151]. In the case of choosing the stimulated saliva collection, it is advisable to use citric acid secretion stimulation at regular intervals of time (e.g., 2 min interval application) [151].

To collect saliva from the sublingual gland only, a similar procedure is used, except that in this case, access to the submandibular gland is blocked in preference to the sublingual gland [151].

Direct Cannulation Methods

One way to collect submandibular or sublingual saliva separately is to cannulate the excretory ducts of the respective glands. However, this procedure is invasive, painful, and requires particular skills [121].

Direct Suction Methods

The simple use of a micropipette suction device [136] or direct syringe aspiration [58] has proven generally successful but with associated drawbacks including frequent partial loss of the saliva sample and some cross-contamination from other salivary glands or from whole saliva.

Customized Collector's Method

Most of this type of devices is based on early proposals such as Schneyer-type segregators [56, 101, 114] or the Block–Brottman saliva collection device [1, 12, 24, 94] with several modifications [1, 24, 44, 56, 94, 101]. These custom-made collectors are normally acrylic fabricated based on an impression of the floor of the mouth taken with dental impression material (e.g., polyvinyl siloxane) with physical separation of sublingual and submaxillary gland ducts and with appropriate tubing to conduct secretions across the collection tubes. A major disadvantage of the customization element of devices is the amount of time and effort needed to construct an individualized collector for each subject and the unavoidable lack of standardization when sampling saliva from different subjects [121].

The Collection of Minor Salivary Gland Saliva

Fluids secreted by the minor salivary glands may be obtained directly from the inner surface of the lips, palate, or buccal mucosa by absorbent paper, pipette [13, 29, 36, 123], or capillary tubes [75, 85]. It should be noted that samples from the minor glands are more viscous in nature and less likely to respond to stimulation than the major gland secretions, so it is overall more difficult to collect [121].

2.3.2   What Is the Best Method to Collect Saliva?

Independently of the method or type of saliva sample chosen, settings should be as standardized as possible for all participants in the study, comprising sampling procedure, processing, and storage conditions. It seems common sense to say it but, the collection method represents most probably the key factor for the successful proteomics analysis of saliva.

First, it is imperious to decide what type of saliva is the target of our study. As was already said, whole saliva is a complex mixture containing everything that is mixed in the oral cavity. Therefore, gland-specific saliva collection is necessary for investigating the pathology or functionality of a specific salivary gland [44, 68], while whole saliva is most frequently used for general studies including a systemic diseases research. An argument against the use of the whole saliva includes contamination with sputum, serum, food debris, and many other non-salivary components [68]. Nevertheless, collection of whole saliva is by far the method most often used. The collection of oral fluids from individual salivary glands seems time-consuming, needs a collecting device, and is rather disagreeable for the individual subjects, whereas the collection of whole saliva is easy to perform, comfortable, inexpensive, and noninvasive, being expected a better collaboration from the study subjects.

When comparing whole unstimulated saliva collection procedures, the passive drool is considered by many researchers to be the gold standard, since it avoids any kind of bias such as reflex stimulation or differentiated contributions of salivary glands, providing the purest sample possible and allows researchers to "biobank" samples for future testing. However, in practice it is not so easy to maintain patients steady for several minutes (in average 10 min), without changing position, coughing, or undergoing some type of stimulation. Probably that is the reason why WHO/YARC (2007) advise the use of the spitting method. Comparing methodologies, as expected and according to Michishige et al. [87], suction strongly stimulates saliva secretion of at least two times if compared with the spitting method. Given the ease of use by the operator, apparent standardization of the procedure and the huge variety of collecting systems that are commercially available, the absorbent method seems to be the most common method for saliva collection in studies. Like the suction method, the absorbent method introduces some degree of stimulation in saliva collection when compared to the unstimulated drooling and spitting methodologies leading to higher saliva flow rates [93] and lower protein concentration [111]. Moreover, the collected saliva volume by different absorbent methods depends on the collection devices and the sampling location in the mouth explaining high individual variances frequently found [93]. When compared, different absorbent methods show significant differences in terms of recovery and storage conditions [85]. In defense of absorbent methods, they filter and help to eliminate cell debris, membranes, protein aggregates, and bacterial cells that may contribute for a better quality of the sample and longer stability at room temperature [111].

For general purposes of saliva analysis, unstimulated whole saliva collected by the passive drool technique is recommended. It is a longstanding method, used at least since the nineteenth century for the analysis of salivary calcium (Ca2+) [150]. Procedures should be kept simple, standardized to get better reproducibility and repeatability on saliva proteomics analysis. Since there are no standard values for salivary constituents it is advisable, whatever the chosen method, to always estimate flow rates and total protein amount.

2.4   Factors That Influence Saliva Composition

Several factors may influence the flow rate and composition of whole saliva resulting in a high interindividual and intraindividual variability. When defining the study population and prior to saliva peptidomics/proteomics, it is important to understand the influence of the variability of human phenotypes and behavior and environment on individual salivary protein signatures.

2.4.1   Aging

Up to 30% of the secretory tissue may be lost with aging, however, with little or no decrease in the stimulated flow rate [86]. It is worth of note that hyposalivation is a frequent observation in elderly due to medication for age-related chronic diseases [99]. Moreover, like most of the physiological functions, the immune activity decreases with age as evidenced by a decline in salivary immunoglobulin concentrations [21, 67]. Age has particularly notorious effects on the salivary proteome pattern in human subjects mostly traduced with a general decrease in the expression of many proteins [43, 92]. For instance, a significant age-associated decrease in histatin concentration for the parotid saliva, as well as for submandibular/sublingual saliva was reported [65]. Nevertheless, other saliva proteins vary their expression with aging such as amylase, whose levels increase up to the middle age (40s) and decreased afterward [67]. Interestingly, it seems that the content of salivary N-glycoproteins increases with age (more markedly in males) mainly the acidic and low molecular weight glycoproteins [130]. Viscosity changes and saliva secreted volume have been reported during aging [92].

2.4.2   Gender

Human body physiology is different in males and females, and gender differences also exist in salivary gland secretion. Lower salivary pH, buffering capacity, protein content [106], and mean salivary flow rate in unstimulated saliva [79] have been reported in female subjects. Bearing in mind that sex steroids are lipophilic, and that it is accepted that approximately 10% of them passively diffuse from plasma to saliva [73] and so, it is expected to exert an influence on saliva of different sexual hormone levels during menstrual cycle. It was reported in parotid saliva that during midcycle there are significantly enhanced concentrations of ionized calcium, total calcium, inorganic phosphate, chloride, and sodium (potassium inversely varied with sodium) with maximal output of total protein during midcycle and menstruation [81].

There are gender differences in the unstimulated salivary proteome mainly associated with immune function, metabolism, and inflammation [43, 156]. Giving some examples for some of the most representative salivary proteins, the salivary kallikrein excretion in the females is higher than in males [80] in particular in females older than 40 years [64], MUC7 and lysozyme activities are higher in females while MUC5B and secretory IgA are lower [106], whereas no sex differences are found for histatins [65], salivary α-amylase [79, 106], albumin, cystatin S, and protease activity [106]. The higher susceptibility of females to Sjögren's syndrome and certain forms of salivary gland cancer probably reflects gender-based differences [73].

2.4.3   Circadian Rhythms

The physiological salivary secretion is modulated by nerve signals by the autonomic nervous system and by the central nervous system. An example of systemic influence on the salivary secretion is the circadian rhythm, which affects salivary flow and saliva composition [105]. Nevertheless, the presence of core clock proteins in the mucous acini and striated ducts of salivary glands suggests an important local role in circadian oscillation of salivary secretion [112].

The circadian rhythms of the salivary flow rate and composition must influence the concept of normal values. And in any study on saliva, the time of day of sampling may have an important impact on the results. Many investigators collect samples at the beginning of the working day, when the unstimulated flow rate and sodium concentrations are showing the most rapid rate of change, since during sleep the flow rate is extremely low [28]. So, it is advisable to only start the collection of saliva in the morning, after a period of complete arousal and stabilization of the salivary secretion (for example of 2h).

Regarding protein expression and circadian rhythms, variations are not equal for all salivary glands, with a strong disparity in the total protein secreted by the parotid during the day with strong influence on the concentration and composition of this type of saliva [28] and a small variation for whole saliva [28, 111]. These differences reflect stability of the submandibular and sublingual gland protein production. Nevertheless, unstimulated whole saliva shows significant circadian rhythms in flow rate and in the concentrations of sodium and chloride [28]. The correlations between salivary proteins and the daytime variations are poorly known. According to Rantonen et al. [108] within-subject variations for several individual salivary proteins and total protein concentrations during day suggest that these proteins are subject to short-term variation at the time of collection.

We observed considerable changes in the specific O-glycan types in human whole saliva during a day, which may be caused by changes in the salivary concentrations of specific proteins or attributed to changes in protein-specific glycosylation profiles [74].

2.4.4 Blood

Glandular function is dependent on local perfusion and thus on the dynamics and changes of the circulatory system. Alteration in the blood perfusion of the salivary glands has impact on the secretory flow and in the process of reabsorption of water and sodium. Changes in saliva secretion may be induced by variations in blood pressure, the use of medication, and several pathological conditions, among others, diabetes, hepatic, and autoimmune diseases [86]. Zhang et al. [161] reported that there was a huge difference on the pattern levels of submandibular gland protein expression for hypertensive rats and specifically found an aquaporin 5 decreased expression and parvalbumin upregulation, which are correlated with water transport and intracellular Ca^{2+} signal transduction and may mechanistically explain how hypertension suppresses saliva secretion.

2.4.5 Drug Effects

A detailed description of the numerous drugs that influence glandular function is beyond the scope of this section. However, some general remarks should be made. To view a list of medications affecting salivary gland function and inducing xerostomia or subjective sialorrhea, please consult a recent excellent review made by Wolff et al. [154].

Many medications can have the following adverse effects: salivary gland dysfunction (SGD), including salivary gland hypofunction (SGH) (objective decrease in salivation) or sialorrhea (objective excessive secretion of saliva), xerostomia (subjective feeling of dry mouth), or subjective sialorrhea (feeling of having too much saliva) [154]. Most drugs that cause salivary gland secretion alterations act on the nervous system, both central and peripheral. Drugs with antagonistic actions on the autonomic receptors since the secretory cells are supplied with muscarinic M1 and M3 receptors, α1 and β1-adrenergic receptors, and certain peptidergic receptors involved in the initiation of salivary secretion cause gland dysfunction and mainly oral dryness [138]. In some cases, the cause of oral dryness is not so evident, as with alendronate that reduces saliva secretion [40]. The number of patients adversely affected by a specific drug and the severity of the effect of that drug are usually dose dependent [2].

Among the 106 medications that have documented evidence of strong or moderate interference on the salivary gland function more than half are used to treat nervous system diseases or have a direct effect on the central nervous system, such is the case of opioids and many drugs from the therapeutic groups of anti-epileptics, anti-Parkinson drugs, psycholeptics (includes many hypnotics and sedatives), and psychoanaleptics (including the most used antidepressants). Another important group are drugs used in cardiac therapy, namely, from the subgroups of antiarrhythmic, antihypertensives, diuretics, beta-blockers, and calcium channel blockers and with less effect agents acting on the renin–angiotensin system. Other drugs belong to alimentary tract and metabolism drugs like antiemetics and antinauseants (e.g., scopolamine/hyoscine), several urological drugs, antineoplastic agents (bevacizumab), bisphosphonates [154], and the majority of antihistaminics [86, 154]. Accelerated flows are seen after the administration of cholinergics (e.g., physostigmine and neostigmine) or parasympathomimetics (e.g., pilocarpine and cevimeline) [86, 154], which are used for the stimulation of salivary flow in patients experiencing dry mouth although the adverse effect profile of these drugs upon systemic administration restricts their use [154]. Apart from age-dependent changes during prolonged drug administration, effects of medication on salivary glands are reversible [86].

It is desirable to monitor changes in saliva, namely, salivary flow rate and composition, after starting the administration of a drug [154] to help investigators in the evaluation of its influence on the population under study.

2.4.6   Other Factors

In the current state of knowledge, it is advisable to take into consideration the construction of the study population and the potential influence of multiple variables that have been identified as potential factors that affect the salivary composition, although no evidence or conflicting results have been presented. Among these factors are body mass index, education, and in particular smoking, which seem to have strong effects on the salivary proteome pattern [92]. Malnutrition in early childhood or situations of continuous nutritional stress significantly reduce saliva flow rates [107]. Recently, ethnic differences in the human plasma proteome have been reported by Cho et al. [25], who found differences in the South Korean male adult whole saliva proteome suggesting an association between several saliva proteins and the top 10 deadliest diseases in South Korea.

Lastly, it should be referred that room conditions may influence the composition of saliva samples. Even a small change in ambient temperature (about 2 °C) in a warm climate may be sufficient to influence unstimulated salivary flow rate with a decrease of salivary flow whenever the ambient temperature increases [70].

2.4.7   Advises to Reduce Variability in Saliva Collection

Although there are no standardized procedures, there are a set of instructions and considerations followed by most authors, regarding the conditions of saliva collection, to ensure the least amount of interference and greater reproducibility of the collected samples.

(i) *To avoid diurnal variation*

Saliva collection should be done in the morning [39, 111], 2 h after waking up, to minimize the influence of circadian rhythms.

(ii) *To avoid changes in the oral environment.*

Saliva collection is recommended to be done preferably in starvation or at least in refrain from eating and drinking for at least 2 h prior to collection, oral hygiene procedures at least 1 h prior to collection; dental treatments should be avoided at least 24 h before collection; a 15 min of rest before collection is mandatory; subjects need to refrain talking and coughing [24].

(iii) *Ensure the homogeneity of the study population.*

Subjects should be observed by a dental professional for oral health evaluation and patient's medication or drug abuse habits documented. Smoking, stress, and medication may induce significant variations into saliva compositions.

(iv) *Ensure the quality of the sample.*

Subjects should rinse their mouth with tap water (e.g., for 30 sec) to remove desquamated epithelial cells, microorganisms, or food and drink remnants [111] and rest for 5 min before collection to avoid sample dilution [39, 57, 151]; blood-contaminated samples must be rejected or identified [151]; during the collection process, the sample tubes should be kept on ice [39, 151].

In order to avoid potential interferences between the analyte and the material of the collection device, it is advisable the use of low-affinity plastic containers [57, 85].

The recipient (normally vials) should be oversized in relation to the sample volume to accommodate the expansion of saliva during freezing.

(v) *Evaluate the flow rate.*

When collecting saliva, the total time necessary to collect is recorded and sample volume measured in order to obtain a secretion rate (output per unit of time, mL/

min). Low flow rates are an indication of salivary gland pathological conditions or of medication (e.g., tricyclic antidepressants), while elevated flow rates will be seen under a number of different conditions such as gingivitis, recent prosthesis, dominant cholinergic activity in Parkinson's disease, or intoxication, among others. The effects are more dramatic in resting saliva on account of intensified water reabsorption in the resting state [86].

(vi) *Latency time*

When collecting stimulated saliva, a latency time elapses between the application of a stimulus and the appearance of saliva, with an interrelation between flow rate and latency time. In healthy glands, a period of about 20 sec is expected, and if values exceed 60 sec, it should be considered pathological [86].

2.5   Sample Preparation

During or immediately after collection, saliva should be placed on ice [1]. This procedure avoids protein degradation for few hours, and without preclearing the degradation is even quicker [41]. In fact, proteome alterations were detected in less than 30 min in untreated samples [113]. The addition of protease inhibitors to whole saliva, but particularly to the clean extract obtained after centrifugation, is mandatory in face of the high proteolytic activity existent in saliva [1]. Endogenous proteases (salivary glands or exfoliating cells) and exogenous proteases (oral flora) contribute to the overall proteolytic activity that occurs post sample collection. Cocktails of protease inhibitors should include PMSF, pepstatin A, leupetin, aprotinin, EDTA, antipain, phenyl methyl sulfonyl fluoride, thimerosal, and/or bestatin E-64 [110]. However, the addition of protease inhibitors might increase the complexity of proteome analysis, particularly when inhibitors are peptides. The addition of sodium azide (NaN3) to saliva specimens to prevent bacterial growth is not recommended once it interferes with proteome analysis. Moreover, saliva specimens should be stored at −80 °C until proteome/peptidome analysis is performed, to avoid posttranslational modifications and protein precipitation, which were reported following sample storage at −20 °C for 3 days. Freezing frequently resulted in significant protein loss, even if quick freezing is used, and even in such conditions proteins are not stable for a longer run [30, 41].

Since saliva is an inhomogeneous liquid with varying viscosity, before analysis, several sample treatments like mixing, dialysis, vortexing, sonication, centrifugation, or ultrafiltration are usually applied. The majority of studies on the characterization of saliva proteome/peptidome start with a centrifugation-based clearance step to remove insoluble material [7]. This procedure is particularly important for the analysis of whole saliva; however, it can lead to the loss of some salivary proteins/peptides, especially when performed after freezing/thawing cycles. So, the centrifugation of saliva specimens should be performed immediately after collection.

The addition of a chaotropic/detergent solution followed by a sonication cycle before the centrifugation step might be advised to promote the disruption of heterotypic complexes such as the ones involving mucins and other proteins. With these experimental steps, the recovery of salivary proteins such as amylase, mucins, cystatins, and histatin is improved. The centrifugation step should be optimized regarding the length and speed applied because some salivary proteins might coprecipitate during centrifugation (e.g., cystatins, PRP, and statherin). Alternatively, centrifugation might occur in tandem with protein precipitation with trichloroacetic acid (TCA) and/or acetone to avoid protein losses [92]. In this case, mucins and other acidic insoluble proteins are disregarded. Instead of being centrifuged, saliva might be filtered using 1.20 μm and 0.45 μm pore size filters, respectively, to remove small particles of food debris and saliva components and, eventually, concentrated with centrifugal filter devices of 3 kDa. This procedure is more time-consuming compared to centrifugation and might lead to the loss of salivary proteins that are retained in the filter. Filtration is always required for the analysis of salivary peptidome. Previously centrifuged or filtered salivary specimens might be used. Filter devices of 30 kDa (from Amicon or Vivaspin) are usually preferred for the separation of peptides. Nevertheless, filters of 10–50 kDa are also used sometimes [146, 147].

When targeting specific classes of salivary proteins or low abundant ones, enrichment strategies should be considered. These strategies usually involve a solid phase matrix (column or beads) with affinity for a given protein modification, such as TiO_2 for phosphorylated proteins or lectins for glycosylated ones. There are several commercially available kits for the enrichment and concentration of specific classes of salivary proteins [42, 47, 122, 125].

2.5.1 Salivary Peptidome

While the term proteomics has been used for high-throughput analysis of proteins expressed by a living system, peptidomics, defines the comprehensive analysis of small peptides and polypeptides of a biological sample (peptidome), less explored or even unexplored by proteomics [3, 4, 83, 133–135, 139]. Thus, efforts have been made in an attempt to characterize salivary peptidome, resulting in up-to-date identification of over 2000 peptides [4].

For peptidome analysis, peptide isolation may be performed passing the supernatant through a sequence of filters, 100 kDa (Centricon 100, Millipore, USA) and 50 kDa and 10 kDa (Vivaspin 500). Addition of agents such as the guanidine 6 M, 3:1 prior to centrifugation or acidified with 0.2% TFA in the proportion of 1:1 and centrifuged at 12000 × g for 30 min (4 °C). The fractions corresponding to the retentate and eluate from the 10-kDa filters from the three methodologies were considered for further analysis [145].

For filter-aided sample preparation (FASP) approach to saliva analysis, use spin filters (30 kDa cutoff) loading saliva samples, and after several washings filtrates (peptides) and retentates (proteins) are analyzed [134]. Unfortunately, FASP is associated with significant peptide loss, a drawback that can be overcome by pre-passivating the filter unit with a detergent, such as Tween 20, which may reduce peptide loss by 300% [38] or by the use of a SDS 1% (w/v) with ammonium bicarbonate wash [134].

2.6 Acquired Enamel Pellicle

Immediately after brushing saliva adsorbs to the tooth forming a pellicle, named enamel pellicle [11] or acquired enamel pellicle (AEP) mainly formed by selective adsorption of proteins and peptides [45, 78]. Pellicle formation is a dynamic process that carries on as a selective process, leading to the formation of two salivary pellicle layers [51].

From previous in vitro studies with hydroxyapatite (HAP) about adsorption and crystallization modification in the presence of a salivary proteins, it is possible to conclude that the ability of a protein segment to bind to HAP surfaces depends on the number and position of the charges, with positive or neutral parts binding less strongly to HAP, whereas segments with several negative charges adsorbing with high affinity illustrated by the greater adsorption of the more acidic statherin and in decreasing order of acidity the acidic proline-rich phosphoproteins (PRP) and cystatins with positively charged histatin 5 or amylase and mucin glycoproteins which lack highly charged segments, adsorbing considerably less [66].

In resume, enamel pellicle is created by the overlap of successive protein layers, starting with a selective and fast adsorption of phosphate- and calcium-binding peptides and small proteins onto the enamel surface [45, 51, 53, 78]. The calcium-binding peptides present in the basal pellicle layer provide a region of high calcium concentration close to the tooth surface and favor calcium exchange between saliva and the tooth surface in an important process for the demineralization/remineralization of the enamel [8, 45, 132]. After the adsorption of low molecular weight proteins, the formation of the salivary pellicle continues by the adsorption of larger salivary proteins and protein aggregates with time, as a coat resulting mainly from protein–protein interactions [51, 53, 78].

A total of 363 proteins were already identified in the AEP collected in vivo [137] although only a minor part corresponds to species secreted by the salivary glands [78, 137]. Typical proteins are statherin, histatins, proline-rich proteins, lactoferrin and cystatins, serum proteins like albumin and immunoglobins, mucins such as MUC5B and MUC7, and several enzymes incorporated in the pellicle in an active conformation like lysozyme, amylase, and peroxidase [55].

Due to its composition, the AEP forms a protective interface between the tooth surface and the oral cavity in a similar process that occurs in all oral tissues giving protection to the mucosa [45], reducing friction and abrasion. AEP also acts as a semipermeable barrier, which modulates the mineralization/demineralization processes and adherence of the microbial flora (mainly bacteria) that forms dental plaque [11, 137, 142, 143, 149].

Significant differences in protein composition and abundance are evident between subjects, thus indicating unique individual pellicle profiles [32]. Many factors can influence salivary film formation, namely, the number of different proteins present, protein size, individual protein concentration, and free ions, through increased/decreased level of electrostatic interaction and protein cross-linking [53]. Moreover, the protein content and the different proportions of calcium and other mineral ions that can influence the ionic strength of whole saliva can modify the protective effect that the salivary pellicle provides against dental erosion [11]. It was already shown that patients with dental erosion display differences in the composition of the salivary pellicles with less total protein, reduced amount of statherin (calcium-binding protein), and reduced amount of calcium [8, 17]. Exogenous proteins from diet, such as casein or ovalbumin, and the incorporation of food polymers have anti-erosive proprieties when incorporated within the pellicle [149].

Another aspect that should be considered is the different composition of AEP according to teeth location. The secretion of the salivary glands differs in protein composition, and the ducts of different glands drain in different mucosal locations; thus the composition of pellicles formed on the various parts of the dentition varies [14]. The AEP is thickest on the lingual surfaces of the lower teeth, since this region is constantly bathed by saliva excreted from the submandibular and sublingual glands [14] and thinnest on the palatal surface of maxillary anterior teeth, because these surfaces are exposed to shear forces from the rubbing action of the tongue and are barely bathed by saliva [6]. Ventura et al. using proteomics, show that the protein profile of the enamel pellicle varies according to its specific location in the dental arches. In this work from the 363 identified proteins only 25 were common to all the locations [137].

2.6.1 Enamel Pellicle Collection

Most of what was said for general procedures on studies with saliva is also valid for enamel pellicle research. In addition, for in vivo studies, dental prophylaxis treatment namely teeth polishing without the use of additive is advisable [164].

2.6.1.1 Methods

Proteomics enamel pellicle studies may be carried out in vitro [8, 52, 66, 119, 140, 159] usually based on incubation of material samples with collected saliva, in situ [143, 144], and in vivo [52, 55, 159, 160].

2.6.1.2 In Vitro Studies

Hydroxyapatite, in the form of powder [66] or discs [119], has largely been studied as a model for the enamel pellicle, establishing base knowledge for the comprehension of the in vivo process. Other materials have also been used such as oxidized silicon surfaces [8], or even mammal's teeth enamel. For in vitro studies and in situ studies, it is possible to use enamel prepared from mammal- extracted teeth (e.g., human molars), by cutting and polishing pieces of enamel surfaces and storing them in a mineral solution [11].

It was concluded that less than a 2 h in situ formed pellicle layer protects the enamel surface to a certain extent against demineralization [54].

2.6.1.3 In Situ Studies

In vitro studies do not mirror the situation in the oral cavity and for that in situ or in vivo approaches are to be preferred. It is argued that in situ studies are preferable since a complete removal of the basal pellicle layer is not achieved by in vivo collecting methods [52, 54, 55]. Enamel plaques mounted on the palatal aspect of removable acrylic splint and exposed to the oral environment can be used [54], but the potential influence on the enamel pellicle from the remaining acrylic monomers resulting from an incomplete polymerization during the manufacture of the device must be considered. Adhesion of an enamel plaque directly on teeth is possible but requires the help from a dental professional [144].

2.6.1.4 In Vivo Studies

For in vivo experiments, the pellicle is scraped off from the dental hard tissue with curettes or wiped down with small sponges [52, 55] or more conveniently with collection strips (electrodes filter paper) [118, 137, 164]. None of these approaches ensure complete removal of the basal pellicle layer [52, 55]; however the use of acids improve the collection yield, for example, using strips pre-dipped in 3% citric acid [118, 137, 164].

The collection should be done from each quadrant in both dental arches after rinsing the teeth with deionized water and drying with compressed air and insulation with cotton rolls. To avoid any contamination from the gingival margin, pellicle collection is made only from the coronal two thirds of the labial/buccal surfaces [118, 137, 164].

To recover pellicle proteins from the collection devices, it is possible to use a solution containing NH_4HCO_3 50 mM pH 7.8 or a solution with 6 M urea, 2 M thiourea [137], that after is submitted to vortex, sonication, and centrifugation to obtain the protein extract in supernatant and to eliminate debris coming from samples and from the collection materials [78, 118, 137, 164]. For the study of the peptidome, it is possible to filter the protein solution by centrifugal filtration using molecular weight cutoff membranes (e.g., 10 kDa) [164].

2.7 Exosomes

Exosomes are 30–100 nm spherical membrane-bound vesicles generated by the endosomal pathway (released through exocytosis of multivesicular bodies (MVBs) and secreted by virtually all cell types and present in many body fluids including saliva. They consist of a lipid bilayer (phospholipids, lipids, polysaccharides, protein receptors) and an inside part which contain lipids, proteins, DNA, and RNA specific to their cell of origin [72, 84, 103, 148]. Furthermore, the molecular composition of the exosomes mirrors a particular physiological status of the producing cell and tissue [128].

Salivary exosomes may arise from the oral mucosa or from each salivary gland that in addition to the "normal" salivary secretion pathways also can secrete exosomes [100]. With a diameter around 47 nm and a density around 1.11 g/mL [62], most of the content of the exosomes in saliva resemble those in plasma [162]. In oral pathological situations, exosomes may show quantitative differences, as in the case of exosomes from patients with oral cancer, which present irregular morphologies, increased vesicle size, and higher intervesicular aggregation [117]. In addition, due to its easy accessibility, saliva has become a potential source for exosomal biomarkers for diagnostic and prognostic assessments of systemic diseases as was shown for pancreatic cancer [77] and for inflammatory bowel disease (ulcerative colitis and Crohn's disease) [163].

2.7.1 Methods

Several proteomics studies of salivary exosomes have been reported with disparity results highlighting that most of the current methods for exosome processing only concentrate exosomes and in reality do not isolate them.

2.7.1.1 Previous Sample Treatments

Previous to exosome isolation, saliva collection is performed as was described in the previous sections.

Although simple clearing sample treatments are applied, the most common procedure is based on a series of differential centrifugations, which first remove cellular debris and contaminants, with a first mild centrifugation of saliva that removes whole cells and debris (e.g., 2600 × g for 15 min at 4 °C) [100], followed by a second centrifugation of the supernatant (e.g., at $12,000 \times g$ for 20 min) that removes residual organelles and large membrane fragments [100].

To improve salivary exosome isolation, several additional procedures have been attempted such as sample filtration [155] to remove amylase [33, 131] and addition of a protease inhibitor cocktail [155]. Notably, saliva is highly viscous in nature and it is very difficult to apply filtration procedures with specific membrane filters before it undergoes any ultracentrifugation [100]. Therefore, to reduce the viscosity of saliva, it may be only diluted (e.g., with PBS 1:1) [155] or disrupted by sonication [62] previously to filtration.

At the end, whatever the salivary exosome isolation procedure chosen, it is necessary to characterize the obtained sample, evaluating the quality and content of the vesicle population obtained (vesicle integrity, size, density, expression of known positive markers) using among others transmission electron microscopy (TEM), flow cytometry (FC), Western blot analysis, LC-MS/MS [104], or AFM [117].

2.7.1.2   Ultracentrifugation

The most common technique for concentrating exosomes in general and also for salivary exosomes is ultracentrifugation [104]. The separation occurs based on size and density with sample ultracentrifugation up to a speed of $200,000 \times g$ and exosomes pelleted from the remaining supernatant [104]. The volume of the samples is normally high, varying between 30 mL [46] and 50 mL [100] of saliva mixed with an equal volume of PBS. To maximize the number of exosomes harvested, a second step of ultracentrifugation may be used [91]. It is considered that this isolation methodology is appropriate for proteomic studies [104]. As an example, parotid saliva exosomes may be isolated [46] following the protocol of Pisitkun et al. [102] for the separation of urine exosomes, which allowed the identification of 491 proteins in the exosome fraction of human parotid saliva. Typical exosome proteins, cytosolic and membrane proteins, comprise the largest category of identified proteins, suggesting that secretion of exosomes by the parotid glands reflects the metabolic and functional status of the gland and may also be useful in the diagnosis of systemic diseases [46].

2.7.1.3   Density Gradient Separation

Although ultracentrifugation is simple and thus widely employed, sample preparations are highly contaminated by other membranous vesicles of different sizes, apoptotic blebs, cellular debris, and large protein aggregates such as proteins non-specifically associated with vesicles. For these reasons, density gradient centrifugation has been considered the "gold standard" for the isolation of exosomes [62].

A variety of density gradient approaches have been described in the literature, using either linear or discontinuous gradients with sucrose [72], OptiPrep [69], or

Percoll [104], which allow separation of exosomes from other types of vesicles or cellular components.

Comparing sucrose, iohexol, and iodixanol for salivary exosome separation, it was concluded that iodixanol yields the best results. Authors propose a density gradient centrifugation isolation protocol for salivary exosomes in which a pretreatment of saliva by sonication and use of iodixanol enables salivary exosomes isolation in a 4 h protocol [62].

2.7.1.4 Other Methods

Other authors separate salivary exosomes with other approaches, namely, gel filtration [96] or immunoaffinity [69]. Two types of extracellular vesicles were separated in human WS by ultrafiltration and gel-exclusion column chromatography [96, 97]. These two kinds of salivary exosomes, with a mean diameter of 83.5 nm and of 40.5 nm as calibrated by transmission electron microscopy, differ not only in size but in protein composition. Proteomic analyses allow to identify a total of 101 and 154 proteins on smaller and larger exosomes, respectively, with 68 common identifications [96]. It was suggested that the heterogeneous structure of salivary exosomes may indicate that exosomes derive from different parts of the salivary glands [96, 97].

2.7.1.5 Commercial Approaches

A commercial chemical-based agent, the ExoQuick™, designed to precipitate exosomes was claimed to be suitable for precipitation of salivary exosomes even from small volumes of saliva; however, authors assume considerably more biological impurities (non-exosomal- related proteins/microvesicles) if compared with ultracentrifugation [165]. This reagent, according to the manufacturer (System Biosciences Inc.), is a polymer that gently precipitates exosomes and microvesicles between 30 and 200 nm. Initially not advised to saliva samples, different authors [158, 165] followed the manufacturer's recommendations and introduced some small modifications and adjusted the kit for saliva. These authors showed that it can be used with saliva volumes higher than 0.5 mL and with an incubation period of 12 h at 4 °C. Briefly, after clearing saliva samples by centrifugation ($3000 \times g$, 15 min), supernatant should be mixed with the reagent, incubated overnight, and centrifuged twice ($1500 \times g$, 30 min + 5 min), and the final pellet should be resuspended with PBS and then kept at −80 °C until further analysis [165].

Although higher exosome quality, with intact morphology, is achieved by ultracentrifugation, density gradient ExoQuick™ precipitation seems to be useful for rapid isolation and increased exosome recovery, but the purity and quality of the sample preparation still need to be confirmed. Furthermore, ExoQuick-TC™ also precipitates other abundant proteins in the sample and cannot preferentially isolate, for example, tumor-derived exosomes. Their application requires a knowledge of positive protein markers present on different vesicles populations (this knowledge still is currently lacking for most sample types) [104].

Other commercial exosome extraction approaches were already used to salivary exosome purification, such as the protocol of the Invitrogen Total Exosome Isolation Kit™, whose successful application was confirmed by electron microscopy [163]. Using proteomics to analyze the exosomes isolated with this procedure, 1408 proteins were identified in salivary exosomes from healthy subjects and 2000 proteins from patients with inflammatory bowel disease (IBD)[163]. Using PureE™ isolation kit (101Bio, CA, USA) for the evaluation of proteome profiles of saliva exosomes, Sun et al. [131] identify 319 proteins with around 80% of saliva proteins shared by serum samples. Coincidently, a panel of 11 cancer-related proteins was detected in exosomes from both the body fluids of lung cancer patients.

2.8 Summary

Considering that most publications in the area address qualitative results, a quantitative approach is needed in the near future and in this regard the published data about saliva are mostly not comparable, because different collection methods were used. Moreover, there is no compound identified in saliva until now that may be used as an internal physiological marker for normalization. Individual data are usually corrected for total protein or flow rate.

All steps of the saliva sampling procedure must be validated for each individual saliva, and the technique used to collect saliva should remain consistent across all individuals and within all samples of each individual [85]. As was said before, whatever method is chosen, its use should be standardized as possible for all participants in the study, from sampling procedure to processing and storage conditions.

References

1. Akintoye SO, Dasso M, Hay DI, Ganeshkumar N, Spielman AI (2002) Partial characterization of a human submandibular/sublingual salivary adhesion-promoting protein. Arch Oral Biol 47(5):337–345
2. Aliko A, Wolff A, Dawes C, Aframian D, Proctor G, Ekstro MJ (2015) World Workshop on Oral Medicine VI: clinical implications of medication-induced salivary gland dysfunction. Oral Surg Oral Med Oral Pathol Oral Radiol 120:185–206
3. Amado FM, Vitorino RM, Domingues PM, Lobo MJ, Duarte JA (2005) Analysis of the human saliva proteome. Expert Rev Proteomics 2(4):521–539. Review
4. Amado F, Lobo MJ, Domingues P, Duarte JA, Vitorino R (2010) Salivary peptidomics. Expert Rev Proteomics 7(5):709–721. https://doi.org/10.1586/epr.10.48. Review
5. Amado FM, Ferreira RP, Vitorino R (2013) One decade of salivary proteomics: current approaches and outstanding challenges. Clin Biochem 46:506–517
6. Amaechi BT, Higham SM, Edgar WM, Milosevic A (1999) Thickness of acquired salivary pellicle as a determinant of the sites of dental erosion. J Dent Res 78(12):1821–1828

7. Aro K, Wei F, Wong DT, Tu M (2017) Saliva liquid biopsy for point-of-care applications. Front Public Health 5:77. https://doi.org/10.3389/fpubh.2017.00077. eCollection 2017
8. Ash A, Ridout MJ, Parker R, Mackie AR, Burnett GR, Wilde PJ (2013) Effect of calcium ions on in vitro pellicle formation from parotid and whole saliva. Colloids Surf B Biointerfaces 102:546–553
9. Atkinson JC (1993) The role of salivary measurements in the diagnosis of salivary autoimmune diseases. Ann N Y Acad Sci 694:238–251
10. Barshir R, Shwartz O, Smoly IY, Yeger-Lotem E (2014) Comparative analysis of human tissue interactomes reveals factors leading to tissue-specific manifestation of hereditary diseases. PLoS Comput Biol 10(6):e1003632. https://doi.org/10.1371/journal.pcbi.1003632. eCollection 2014 Jun
11. Baumann T, Bereiter R, Lussi A, Carvalho TS (2017) The effect of different salivary calcium concentrations on the erosion protection conferred by the salivary pellicle. Sci Rep 7(1):12999. https://doi.org/10.1038/s41598-017-13367-3
12. Block PL, Brottman BS (1962) A method of submaxillary saliva collection without cannulation. NY State Dent J 28:116–118
13. Boros I, Keszler P, Zelles T (1999) Study of saliva secretion and the salivary fluoride concentration of the human minor labial glands by a new method. Arch Oral Biol 44(Suppl 1):S59–S62
14. Carlén A, Börjesson AC, Nikdel K, Olsson J (1998) Composition of pellicles formed in vivo on tooth surfaces in different parts of the dentition, and in vitro on hydroxyapatite. Caries Res 32(6):447–455
15. Carlson AJ, Crittenden AL (1910) The relationship of ptyalin concentration to the diet and the rate of secretion of saliva. Am J Phys 26:169–177
16. Carpenter GH (2013) The secretion, components, and properties of saliva. Annu Rev Food Sci Technol 4:267–276. https://doi.org/10.1146/annurev-food-030212-182700
17. Carpenter G, Cotroneo E, Moazzez R, Rojas-Serrano M, Donaldson N, Austin R, Zaidel L, Bartlett D, Proctor G (2014) Composition of enamel pellicle from dental erosion patients. Caries Res 48(5):361–367. https://doi.org/10.1159/000356973. Epub 2014 Mar 6
18. Caseiro A, Vitorino R, Barros AS, Ferreira R, Calheiros-Lobo MJ, Carvalho D, Duarte JA, Amado F (2012) Salivary peptidome in type 1 diabetes mellitus. Biomed Chromatogr 26(5):571–582. https://doi.org/10.1002/bmc.1677. Epub 2011 Sep 6
19. Caseiro A, Ferreira R, Padrão A, Quintaneiro C, Pereira A, Marinheiro R, Vitorino R, Amado F (2013) Salivary proteome and peptidome profiling in type 1 diabetes mellitus using a quantitative approach. J Proteome Res 12(4):1700–1709. https://doi.org/10.1021/pr3010343. Epub 2013 Feb 25
20. Castagnola M, Picciotti PM, Messana I, Fanali C, Fiorita A, Cabras T, Calò L, Pisano E, Passali GC, Iavarone F, Paludetti G, Scarano E (2011) Potential applications of human saliva as diagnostic fluid. Acta Otorhinolaryngol Ital 31(6):347–357
21. Challacombe SJ, Percival RS, Marsh PD (1995) Age-related changes in immunoglobulin isotypes in whole and parotid saliva and serum in healthy individuals. Oral Microbiol Immunol 10(4):202–207
22. Chen DX, Schwartz PE, Li FQ (1990) Saliva and serum CA 125 assays for detecting malignant ovarian tumors. Obstet Gynecol 75(4):701–704
23. Chiappin S, Antonelli G, Gatti R, De Palo EF (2007) Saliva specimen: a new laboratory tool for diagnostic and basic investigation. Clin Chim Acta 383(1–2):30–40. Epub 2007 Apr 25
24. Ching KH, Burbelo PD, Gonzalez-Begne M, Roberts ME, Coca A, Sanz I, Iadarola MJ (2011) Salivary anti-Ro60 and anti-Ro52 antibody profiles to diagnose Sjogren's Syndrome. J Dent Res 90(4):445–449. https://doi.org/10.1177/0022034510390811. Epub 2011 Jan 6
25. Cho HR, Kim HS, Park JS, Park SC, Kim KP, Wood TD, Choi YS (2017) Construction and characterization of the Korean whole saliva proteome to determine ethnic differences in human saliva proteome. PLoS One 12(7):e0181765. https://doi.org/10.1371/journal.pone.0181765. eCollection 2017

26. Chojnowska S, Baran T, Wilińska I, Sienicka P, Cabaj-Wiater I, Knaś M (2017) Human saliva as a diagnostic material. Adv Med Sci 63(1):185–191. https://doi.org/10.1016/j.advms.2017.11.002. Review

27. Crouch DJ, Walsh JM, Flegel R, Cangianelli L, Baudys J, Atkins R (2005) An evaluation of selected oral fluid point-of-collection drug-testing devices. J Anal Toxicol 29(4):244–248

28. Dawes C (1972) Circadian rhythms in human salivary flow rate and composition. J Physiol 220(3):529–545

29. Dawes C, Wood CM (1973) The contribution of oral minor mucous gland secretions to the volume of whole saliva in man. Arch Oral Biol 18(3):337–342

30. de Jong EP, van Riper SK, Koopmeiners JS, Carlis JV, Griffin TJ (2011) Sample collection and handling considerations for peptidomic studies in whole saliva; implications for biomarker discovery. Clin Chim Acta 412(23–24):2284–2288. https://doi.org/10.1016/j.cca.2011.08.023. Epub 2011 Aug 24

31. Delaleu N, Mydel P, Kwee I, Brun JG, Jonsson MV, Jonsson R (2015) High fidelity between saliva proteomics and the biologic state of salivary glands defines biomarker signatures for primary Sjögren's syndrome. Arthritis Rheumatol 67(4):1084–1095. https://doi.org/10.1002/art.39015

32. Delius J, Trautmann S, Médard G, Kuster B, Hannig M, Hofmann T (2017) Label-free quantitative proteome analysis of the surface-bound salivary pellicle. Colloids Surf B Biointerfaces 152:68–76. https://doi.org/10.1016/j.colsurfb.2017.01.005. Epub 2017 Jan 6

33. Deutsch O, Fleissig Y, Zaks B, Krief G, Aframian DJ, Palmon A (2008) An approach to remove alpha amylase for proteomic analysis of low abundance biomarkers in human saliva. Electrophoresis 29(20):4150–4157. https://doi.org/10.1002/elps.200800207

34. Dodds MW, Johnson DA, Yeh CK (2005) Health benefits of saliva: a review. J Dent 33(3):223–233. Epub 2004 Dec 19

35. Dominy SS, Brown JN, Ryder MI, Gritsenko M, Jacobs JM, Smith RD (2014) Proteomic analysis of saliva in HIV-positive heroin addicts reveals proteins correlated with cognition. PLoS One 9(4):e89366. https://doi.org/10.1371/journal.pone.0089366. eCollection 2014

36. Eliasson L, Carlén A (2010) An update on minor salivary gland secretions. Eur J Oral Sci 118(5):435–442. https://doi.org/10.1111/j.1600-0722.2010.00766.x. Epub 2010 Aug 24

37. Elzek MA, Rodland KD (2015) Proteomics of ovarian cancer: functional insights and clinical applications. Cancer Metastasis Rev 34(1):83–96. https://doi.org/10.1007/s10555-014-9547-8. Review

38. Erde J, Loo RR, Loo JA (2014) Enhanced FASP (eFASP) to increase proteome coverage and sample recovery for quantitative proteomic experiments. J Proteome Res 13(4):1885–1895. https://doi.org/10.1021/pr4010019. Epub 2014 Mar 6

39. Esser D, Alvarez-Llamas G, de Vries MP, Weening D, Vonk RJ, Roelofsen H (2008) Sample stability and protein composition of saliva: implications for its use as a diagnostic fluid. Biomark Insights 3:25–27

40. Eviö S, Tarkkila L, Sorsa T, Furuholm J, Välimäki MJ, Ylikorkala O, Tiitinen A, Meurman JH (2006) Effects of alendronate and hormone replacement therapy, alone and in combination, on saliva, periodontal conditions and gingival crevicular fluid matrix metalloproteinase-8 levels in women with osteoporosis. Oral Dis 12(2):187–193

41. Fábián TK, Fejérdy P, Csermely P (2008) Salivary genomics, Transcriptomics and proteomics: the emerging concept of the Oral ecosystem and their use in the early diagnosis of cancer and other diseases. Curr Genomics 9(1):11–21. https://doi.org/10.2174/138920208783884900

42. Ferreira JA, Daniel-da-Silva AL, Alves RM, Duarte D, Vieira I, Santos LL, Vitorino R, Amado F (2011) Synthesis and optimization of lectin functionalized nanoprobes for the selective recovery of glycoproteins from human body fluids. Anal Chem 83(18):7035–7043. https://doi.org/10.1021/ac200916j. Epub 2011 Aug 26

43. Fleissig Y, Reichenberg E, Redlich M, Zaks B, Deutsch O, Aframian DJ, Palmon A (2010) Comparative proteomic analysis of human oral fluids according to gender and age. Oral Dis 16(8):831–838. https://doi.org/10.1111/j.1601-0825.2010.01696.x

44. Fox PC, van der Ven PF, Sonies BC, Weiffenbach JM, Baum BJ (1985) Xerostomia: evaluation of a symptom with increasing significance. J Am Dent Assoc 110(4):519–525
45. Gibbins HL, Yakubov GE, Proctor GB, Wilson S, Carpenter GH (2014) What interactions drive the salivary mucosal pellicle formation? Colloids Surf B Biointerfaces 120:184–192. https://doi.org/10.1016/j.colsurfb.2014.05.020. Epub 2014 May 23
46. Gonzalez-Begne M, Lu B, Han X, Hagen FK, Hand AR, Melvin JE, Yates JR (2009) Proteomic analysis of human parotid gland exosomes by multidimensional protein identification technology (MudPIT). J Proteome Res 8(3):1304–1314. https://doi.org/10.1021/pr800658c
47. Gonzalez-Begne M, Lu B, Liao L, Xu T, Bedi G, Melvin JE, Yates JR 3rd. (2011) Characterization of the human submandibular/sublingual saliva glycoproteome using lectin affinity chromatography coupled to multidimensional protein identification technology. J Proteome Res 10(11):5031–5046. https://doi.org/10.1021/pr200505t. Epub 2011 Oct 13
48. Granger DA, KivlighaEn KT, el-Sheikh M, Gordis EB, Stroud LR (2007) Salivary alpha-amylase in biobehavioral research: recent developments and applications. Ann N Y Acad Sci 1098:122–144. Epub 2007 Mar 1
49. Gröschl M, Rauh M (2006) Influence of commercial collection devices for saliva on the reliability of salivary steroids analysis. Steroids 71:1097–1100
50. Guzman YA, Sakellari D, Arsenakis M, Floudas CA (2014) Proteomics for the discovery of biomarkers and diagnosis of periodontitis: a critical review. Expert Rev Proteomics 11(1):31–41. https://doi.org/10.1586/14789450.2014.864953. Epub 2013 Nov 26. Review
51. Hannig M (1999) Ultrastructural investigation of pellicle morphogenesis at two different intraoral sites during a 24-h period. Clin Oral Investig 3(2):88–95
52. Hannig C, Hannig M (2009) The oral cavity--a key system to understand substratum-dependent bioadhesion on solid surfaces in man. Clin Oral Investig 13(2):123–139. https://doi.org/10.1007/s00784-008-0243-3. Epub 2009 Jan 10
53. Hannig M, Hannig C (2014) The pellicle and erosion. Monogr Oral Sci 25:206–214
54. Hannig M, Hess NJ, Hoth-Hannig W, De Vrese M (2003) Influence of salivary pellicle formation time on enamel demineralization--an in situ pilot study. Clin Oral Investig 7(3):158–161. Epub 2003 Jul 26
55. Hannig C, Hannig M, Kensche A, Carpenter G (2017) The mucosal pellicle - an underestimated factor in oral physiology. Arch Oral Biol 80:144–152. https://doi.org/10.1016/j.archoralbio.2017.04.001. Epub 2017 Apr 8
56. Henriques BI, Chauncey HH (1961) A modified method for the collection of human submaxillary and sublingual saliva. Oral Surg Oral Med Oral Pathol 14:1124–1129
57. Henson BS, Wong DT (2010) Collection, storage, and processing of saliva samples for downstream molecular applications. Methods Mol Biol 666:21–30. https://doi.org/10.1007/978-1-60761-820-1_2
58. Hu S, Wang J, Meijer J, Ieong S, Xie Y, Yu T, Zhou H, Henry S, Vissink A, Pijpe J, Kallenberg C, Elashoff D, Loo JA, Wong DT (2007) Salivary proteomic and genomic biomarkers for primary Sjögren's syndrome. Arthritis Rheum 56(11):3588–3600
59. Hu S, Arellano M, Boontheung P et al (2008) Salivary proteomics for oral cancer biomarker discovery. Clin Cancer Res 14(19):6246–6252
60. Il S, Jr P, Chauncey H (1962) Modified Carlson-Crittenden device for the collection of parotid fluid. J Dent Res 41:778–783
61. Isabel Padrão A, Ferreira R, Vitorino R, Amado F (2012) Proteome-base biomarkers in diabetes mellitus: progress on biofluids' protein profiling using mass spectrometry. Proteomics Clin Appl 6(9–10):447–466. https://doi.org/10.1002/prca.201200044
62. Iwai K, Minamisawa T, Suga K, Yajima Y, Shiba K (2016) Isolation of human salivary extracellular vesicles by iodixanol density gradient ultracentrifugation and their characterizations. J Extracell Vesicles 5:30829. https://doi.org/10.3402/jev.v5.30829. eCollection 2016
63. Jasim H, Olausson P, Hedenberg-Magnusson B, Ernberg M, Ghafouri B (2016) The proteomic profile of whole and glandular saliva in healthy pain-free subjects. Sci Rep 6:39073. https://doi.org/10.1038/srep39073

64. Jenzano JW, Hogan SL, Lundblad RL (1992) The influence of age, sex and race on salivary kallikrein levels in human mixed saliva. Agents Actions 35(1–2):29–33
65. Johnson DA, Yeh CK, Dodds MW (2000) Effect of donor age on the concentrations of histatins in human parotid and submandibular/sublingual saliva. Arch Oral Biol 45(9):731–740
66. Johnsson M, Levine MJ, Nancollas GH (1993) Hydroxyapatite binding domains in salivary proteins. Crit Rev Oral Biol Med 4(3–4):371–378
67. Kalipatnapu P, Kelly RH, Rao KN, van Thiel DH (1983) Salivary composition: effects of age and sex. Acta Medica Port 4(7–8):327–330
68. Kalk WW, Vissink A, Spijkervet FK, Bootsma H, Kallenberg CG, Nieuw Amerongen AV (2001) Sialometry and sialochemistry: diagnostic tools for Sjögren's syndrome. Ann Rheum Dis 60(12):1110–1116
69. Kalra H, Adda CG, Liem M, Ang CS, Mechler A, Simpson RJ, Hulett MD, Mathivanan S (2013) Comparative proteomics evaluation of plasma exosome isolation techniques and assessment of the stability of exosomes in normal human blood plasma. Proteomics 13(22):3354–3364. https://doi.org/10.1002/pmic.201300282. Epub 2013 Oct 18
70. Kariyawasam AP, Dawes C (2005) A circannual rhythm in unstimulated salivary flow rate when the ambient temperature varies by only about 2 degrees C. Arch Oral Biol 50(10):919–922. Epub 2005 Apr 7
71. Kaufman E, Lamster IB (2002) The diagnostic applications of saliva--a review. Crit Rev Oral Biol Med 13(2):197–212
72. Keller S, Ridinger J, Rupp AK, Janssen JW, Altevogt P (2011) Body fluid derived exosomes as a novel template for clinical diagnostics. J Transl Med 9:86. https://doi.org/10.1186/1479-5876-9-86
73. Konttinen YT, Stegaev V, Mackiewicz Z, Porola P, Hänninen A, Szodoray P (2010) Salivary glands - "an unisex organ"? Oral Dis 16(7):577–585. https://doi.org/10.1111/j.1601-0825.2010.01669.x
74. Kozak RP, Urbanowicz PA, Punyadeera C, Reiding KR, Jansen BC, Royle L, Spencer DI, Fernandes DL, Wuhrer M (2016) Variation of human salivary O-Glycome. PLoS One 11(9):e0162824. https://doi.org/10.1371/journal.pone.0162824. eCollection 2016
75. Kutscher AH, Mandel ID, Zegarelli EV, Denning C, Eriv A, Ruiz L, Ellegood K, Phalen J (1967) A technique for collecting the secretion of minor salivary glands: I. use of capillary tubes. J Oral Ther Pharmacol 3(5):391–392
76. Lashley KS (1916) The human salivary reflex and its use in psychology. Psychol Rev 23:446–464
77. Lau C, Kim Y, Chia D, Spielmann N, Eibl G, Elashoff D, Wei F, Lin YL, Moro A, Grogan T, Chiang S, Feinstein E, Schafer C, Farrell J, Wong DT (2013) Role of pancreatic cancer-derived exosomes in salivary biomarker development. J Biol Chem 288(37):26888–26897. https://doi.org/10.1074/jbc.M113.452458. Epub 2013 Jul 23
78. Lee YH, Zimmerman JN, Custodio W, Xiao Y, Basiri T, Hatibovic-Kofman S, Siqueira WL (2013) Proteomic evaluation of acquired enamel pellicle during in vivo formation. PLoS One 8(7):e67919. https://doi.org/10.1371/journal.pone.0067919. Print 2013
79. Li-Hui W, Chuan-Quan L, Long Y, Ru-Liu L, Long-Hui C, Wei-Wen C (2016) Gender differences in the saliva of young healthy subjects before and after citric acid stimulation. Clin Chim Acta 460:142–145. https://doi.org/10.1016/j.cca.2016.06.040. Epub 2016 Jun 30
80. Maier H, Menstell S (1986) Influence of sex and age on kallikrein excretion in stimulated human parotid saliva. Arch Otorhinolaryngol 243(2):138–140
81. Maier H, Geissler M, Heidland A, Schindler JG, Wigand ME (1979) [The influence of menstruation cycle on human parotid saliva composition (author's transl)].[Article in German] Laryngol Rhinol Otol (Stuttg) 58(9):706–710
82. Malamud D, Rodriguez-Chavez IR (2011) Saliva as a diagnostic fluid. Dent Clin N Am 55(1):159–178
83. Manconi B, Liori B, Cabras T, Iavarone F, Manni A, Messana I, Castagnola M, Olianas A (2017) Top-down HPLC-ESI-MS proteomic analysis of saliva of edentulous subjects evidenced high levels of cystatin A, cystatin B and SPRR3. Arch Oral Biol 77:68–74. https://doi.org/10.1016/j.archoralbio.2017.01.021. Epub 2017 Jan 31

84. Mathivanan S, Ji H, Simpson RJ (2010) Exosomes: extracellular organelles important in intercellular communication. J Proteome 73(10):1907–1920. doi: https://doi.org/10.1016/j.jprot.2010.06.006. Epub 2010 Jul 1. Review

85. Michalke B, Rossbach B, Göen T, Schäferhenrich A, Scherer G (2015) Saliva as a matrix for human biomonitoring in occupational and environmental medicine. Int Arch Occup Environ Health 88(1):1–44. https://doi.org/10.1007/s00420-014-0938-5. Epub 2014 Mar 12

86. Michels LFE (1991) In: Graamans K, van den Akker HP (eds) Diagnosis of salivary gland disorders. Kluwer Academic Publishers, the Netherlands, pp 139–161

87. Michishige F, Kanno K, Yoshinaga S, Hinode D, Takehisa Y, Yasuoka S (2006) Effect of saliva collection method on the concentration of protein components in saliva. J Med Investig 53(1–2):140–146

88. Milioli HH, Santos Sousa K, Kaviski R, Dos Santos Oliveira NC, De Andrade UC, De Lima RS, Cavalli IJ, De Souza Fonseca Ribeiro EM (2015) Comparative proteomics of primary breast carcinomas and lymph node metastases outlining markers of tumor invasion. Cancer Genomics Proteomics 12(2):89–101

89. Miller CS, Foley JD, Bailey AL, Campell CL, Humphries RL, Christodoulides N, Floriano PN, Simmons G, Bhagwandin B, Jacobson JW, Redding SW, Ebersole JL, McDevitt JT (2010) Current developments in salivary diagnostics. Biomark Med 4(1):171–189

90. Muddugangadhar BC, Sangur R, Rudraprasad IV, Nandeeshwar DB, Kumar BH (2015) A clinical study to compare between resting and stimulated whole salivary flow rate and pH before and after complete denture placement in different age groups. J Indian Prosthodont Soc 15(4):356–366. https://doi.org/10.4103/0972-4052.164907

91. Mueller SK, Nocera AL, Bleier BS (2017) Exosome function in aerodigestive mucosa. Nanomedicine 14(2):269–277. https://doi.org/10.1016/j.nano.2017.10.008

92. Murr A, Pink C, Hammer E, Michalik S, Dhople VM, Holtfreter B, Völker U, Kocher T, Gesell SM (2017) Cross-sectional Association of Salivary Proteins with age, sex, body mass index, smoking, and education. J Proteome Res 16(6):2273–2281. https://doi.org/10.1021/acs.jproteome.7b00133. Epub 2017 May 18

93. Navazesh M, Christensen CM (1982) A comparison of whole mouth resting and stimulated salivary measurement procedures. J Dent Res 61(10):1158–1162

94. Nederfors T, Dahlöf C (1993) A modified device for collection and flow-rate measurement of submandibular-sublingual saliva. Scand J Dent Res 101(4):210–214

95. Nunes LA, Mussavira S, Bindhu OS (2015) Clinical and diagnostic utility of saliva as a non-invasive diagnostic fluid: a systematic review. Biochem Med (Zagreb) 25(2):177–192. https://doi.org/10.11613/BM.2015.018. eCollection 2015. Review

96. Ogawa Y, Miura Y, Harazono A, Kanai-Azuma M, Akimoto Y, Kawakami H, Yamaguchi T, Toda T, Endo T, Tsubuki M, Yanoshita R (2011) Proteomic analysis of two types of exosomes in human whole saliva. Biol Pharm Bull 34(1):13–23

97. Ogawa Y, Tsujimoto M, Yanoshita R (2016) Next-generation sequencing of protein-coding and Long non-protein-coding RNAs in two types of Exosomes derived from human whole saliva. Biol Pharm Bull 39(9):1496–1507. https://doi.org/10.1248/bpb.b16-00297

98. Ohyama K, Baba M, Tamai M, Aibara N, Ichinose K, Kishikawa N, Kawakami A, Kuroda N (2015) Proteomic profiling of antigens in circulating immune complexes associated with each of seven autoimmune diseases. Clin Biochem 48(3):181–185. https://doi.org/10.1016/j.clinbiochem.2014.11.008. Epub 2014 Nov 29

99. Pajukoski H, Meurman JH, Snellman-Gröhn S, Keinänen S, Sulkava R (1997) Salivary flow and composition in elderly patients referred to an acute care geriatric ward. Oral Surg Oral Med Oral Pathol Oral Radiol Endod 84(3):265–271

100. Palanisamy V, Sharma S, Deshpande A, Zhou H, Gimzewski J, Wong DT (2010) Nanostructural and transcriptomic analyses of human saliva derived exosomes. PLoS One 5(1):e8577. https://doi.org/10.1371/journal.pone.0008577

101. Parr GR, Bustos-Valdes SE (1984) A modified segregator for collection of human submandibular and sublingual saliva. Arch Oral Biol 29(1):69–71

102. Pisitkun T, Shen RF, Knepper MA (2004) Identification and proteomic profiling of exosomes in human urine. Proc Natl Acad Sci U S A 101(36):13368–13373. Epub 2004 Aug 23
103. Pisitkun T, Johnstone R, Knepper MA (2006) Discovery of urinary biomarkers. Mol Cell Proteomics 5(10):1760–1771. Epub 2006 Jul 12. Review
104. Principe S, Hui AB, Bruce J, Sinha A, Liu FF, Kislinger T (2013) Tumor-derived exosomes and microvesicles in head and neck cancer: implications for tumor biology and biomarker discovery. Proteomics 13(10–11):1608–1623. https://doi.org/10.1002/pmic.201200533
105. Proctor GB (2016) The physiology of salivary secretion. Periodontol 2000 70(1):11–25. https://doi.org/10.1111/prd.12116
106. Prodan A, Brand HS, Ligtenberg AJ, Imangaliyev S, Tsivtsivadze E, van der Weijden F, Crielaard W, Keijser BJ, Veerman EC (2015) Interindividual variation, correlations, and sex-related differences in the salivary biochemistry of young healthy adults. Eur J Oral Sci 123(3):149–157. https://doi.org/10.1111/eos.12182. Epub 2015 Mar 23
107. Psoter WJ, Spielman AL, Gebrian B, St Jean R, Katz RV (2008) Effect of childhood malnutrition on salivary flow and pH. Arch Oral Biol 53(3):231–237. Epub 2007 Nov 5
108. Rantonen PJ, Meurman JH (2000) Correlations between total protein, lysozyme, immunoglobulins, amylase, and albumin in stimulated whole saliva during daytime. Acta Odontol Scand 58(4):160–165
109. Rao PV, Reddy AP, Lu X et al (2009) Proteomic identification of salivary biomarkers of type-2 diabetes. J Proteome Res 8(1):239–245
110. Robinson S, Niles RK, Witkowska HE, Rittenbach KJ, Nichols RJ, Sargent JA, Dixon SE, Prakobphol A, Hall SC, Fisher SJ, Hardt M (2008) A mass spectrometry-based strategy for detecting and characterizing endogenous proteinase activities in complex biological samples. Proteomics 8(3):435–445. https://doi.org/10.1002/pmic.200700680
111. Rosa N, Marques J, Esteves E, Fernandes M, Mendes VM, Afonso A, Dias S, Pereira JP, Manadas B, Correia MJ, Barros M (2016) Protein quality assessment on saliva samples for biobanking purposes. Biopreserv Biobank 14(4):289–297. https://doi.org/10.1089/bio.2015.0054. Epub 2016 Mar 3
112. Satou R, Sato M, Kimura M, Ishizuka Y, Tazaki M, Sugihara N, Shibukawa Y (2017) Temporal expression patterns of clock genes and aquaporin 5/Anoctamin 1 in rat submandibular gland cells. Front Physiol 8:320. https://doi.org/10.3389/fphys.2017.00320. eCollection 2017
113. Schipper R, Loof A, de Groot J, Harthoorn L, Dransfield E, van Heerde W (2007) SELDI-TOF-MS of saliva: methodology and pre-treatment effects. J Chromatogr B Analyt Technol Biomed Life Sci 847(1):45–53. Epub 2006 Oct 27
114. Schneyer LH (1955) Method for the collection of separate submaxillary and sublingual Salivas in man. J Dent Res 34(2):257–261
115. Schulz BL, Cooper-White J, Punyadeera CK (2013) Saliva proteome research: current status and future outlook. Crit Rev Biotechnol 33(3):246–259. https://doi.org/10.3109/07388551.2012.687361. Epub 2012 May 21
116. Shah P, Wang X, Yang W, Toghi Eshghi S, Sun S, Hoti N, Chen L, Yang S, Pasay J, Rubin A, Zhang H (2015) Integrated proteomic and Glycoproteomic analyses of prostate Cancer cells reveal glycoprotein alteration in protein abundance and glycosylation. Mol Cell Proteomics 14(10):2753–2763. https://doi.org/10.1074/mcp.M115.047928. Epub 2015 Aug 9
117. Sharma S, Gillespie BM, Palanisamy V, Gimzewski JK (2011) Quantitative nanostructural and single-molecule force spectroscopy biomolecular analysis of human-saliva-derived exosomes. Langmuir 27(23):14394–14400. https://doi.org/10.1021/la2038763. Epub 2011 Nov 9
118. Siqueira WL, Zhang W, Helmerhorst EJ, Gygi SP, Oppenheim FG (2007) Identification of protein components in in vivo human acquired enamel pellicle using LC-ESI-MS/MS. J Proteome Res 6(6):2152–2160. Epub 2007 Apr 21
119. Siqueira WL, Bakkal M, Xiao Y, Sutton JN, Mendes FM (2012) Quantitative proteomic analysis of the effect of fluoride on the acquired enamel pellicle. PLoS One 7(8):e42204. https://doi.org/10.1371/journal.pone.0042204. Epub 2012 Aug 1
120. Slowey PD (2013) Commercial saliva collections tools. J Calif Dent Assoc 41(2):97–99, 102-5

121. Slowey PD (2015) Saliva collection devices and diagnostic platforms. In: Streckfus C (ed) Advances in salivary diagnostics. Springer-Verlag, Heidelberg
122. Sondej M, Denny PA, Xie Y, Ramachandran P, Si Y, Takashima J, Shi W, Wong DT, Loo JA, Denny PC (2009) Glycoprofiling of the human salivary proteome. Clin Proteomics 5(1):52–68
123. Sonesson M, Hamberg K, Wallengren ML, Matsson L, Ericson D (2011) Salivary IgA in minor-gland saliva of children, adolescents, and young adults. Eur J Oral Sci 119(1):15–20. https://doi.org/10.1111/j.1600-0722.2010.00794.x
124. Stokes JR, Davies GA (2007) Viscoelasticity of human whole saliva collected after acid and mechanical stimulation. Biorheology 44(3):141–160
125. Stone MD, Chen X, McGowan T, Bandhakavi S, Cheng B, Rhodus NL, Griffin TJ (2011) Large-scale phosphoproteomics analysis of whole saliva reveals a distinct phosphorylation pattern. J Proteome Res 10(4):1728–1736. https://doi.org/10.1021/pr1010247. Epub 2011 Mar 1
126. Streckfus C, Bigler L (2005) The use of soluble, salivary c-erbB-2 for the detection and post-operative follow-up of breast cancer in women: the results of a five-year translational research study. Adv Dent Res 18(1):17–24. [PubMed]
127. Streckfus C, Bigler L, Tucci M, Thigpen JT (2000) A preliminary study of CA15-3, c-erbB-2, epidermal growth factor receptor, cathepsin-D, and p53 in saliva among women with breast carcinoma. Cancer Investig 18(2):101–109. [PubMed]
128. Stremersch S, De Smedt SC, Raemdonck K (2016) Therapeutic and diagnostic applications of extracellular vesicles. J Control Release 244(Pt B):167–183. https://doi.org/10.1016/j.jconrel.2016.07.054. Epub 2016 Aug 2
129. Stuani VT, Rubira CM, Sant'Ana AC, Santos PS (2017) Salivary biomarkers as tools for oral squamous cell carcinoma diagnosis: a systematic review. Head Neck 39(4):797–811. https://doi.org/10.1002/hed.24650. Epub 2016 Nov 29
130. Sun S, Zhao F, Wang Q, Zhong Y, Cai T, Wu P, Yang F, Li Z (2014) Analysis of age and gender associated N-glycoproteome in human whole saliva. Clin Proteomics 11(1):25. https://doi.org/10.1186/1559-0275-11-25. eCollection 2014
131. Sun Y, Liu S, Qiao Z, Shang Z, Xia Z, Niu X, Qian L, Zhang Y, Fan L, Cao CX, Xiao H (2017) Systematic comparison of exosomal proteomes from human saliva and serum for the detection of lung cancer. Anal Chim Acta 982:84–95. https://doi.org/10.1016/j.aca.2017.06.005. Epub 2017 Jun 17
132. ten Cate JM, Featherstone JD (1991) Mechanistic aspects of the interactions between fluoride and dental enamel. Crit Rev Oral Biol Med 2(3):283–296. Review
133. Trindade F, Amado F, Oliveira-Silva RP, Daniel-da-Silva AL, Ferreira R, Klein J, Faria-Almeida R, Gomes PS, Vitorino R (2015) Toward the definition of a peptidome signature and protease profile in chronic periodontitis. Proteomics Clin Appl 9(9–10):917–927. https://doi.org/10.1002/prca.201400191. Epub 2015 May 8
134. Trindade F, Amado F, Gomes PS, Vitorino R (2015) endoProteoFASP: a novel FASP approach to profile salivary peptidome and disclose salivary proteases. Talanta 132:486–493
135. Trindade F, Amado F, Pinto da Costa J, Ferreira R, Maia C, Henriques I, Colaço B, Vitorino R (2015) Salivary peptidomic as a tool to disclose new potential antimicrobial peptides. J Proteome 115:49–57. https://doi.org/10.1016/j.jprot.2014.12.004. Epub 2014 Dec 20
136. Tylenda CA, Ship JA, Fox PC, Baum BJ (1988) Evaluation of submandibular salivary flow rate in different age groups. J Dent Res 67(9):1225–1228
137. Ventura TMDS, Cassiano LPS, Souza E, Silva CM, Taira EA, Leite AL, Rios D, Buzalaf MAR (2017) The proteomic profile of the acquired enamel pellicle according to its location in the dental arches. Arch Oral Biol 79:20–29. https://doi.org/10.1016/j.archoralbio.2017.03.001. Epub 2017 Mar 3
138. Villa A, Wolff A, Narayana N, Dawes C, Aframian DJ, Lynge Pedersen AM et al (2016) World workshop on Oral medicine VI: a systematic review of medication-induced salivary gland dysfunction. Oral Dis 22:365–382

139. Vitorino R (2018) Digging deep into peptidomics applied to body fluids. Proteomics 18(2). https://doi.org/10.1002/pmic.201700401. Review

140. Vitorino R, Lobo MJ, Duarte J, Ferrer-Correia AJ, Tomer KB, Dubin JR, Domingues PM, Amado FM (2004) In vitro hydroxyapatite adsorbed salivary proteins. Biochem Biophys Res Commun 320(2):342–346

141. Vitorino R, Lobo MJ, Duarte JR, Ferrer-Correia AJ, Domingues PM, Amado FM (2005) The role of salivary peptides in dental caries. Biomed Chromatogr 19(3):214–222

142. Vitorino R, de Morais GS, Ferreira R, Lobo MJ, Duarte J, Ferrer-Correia AJ, Tomer KB, Domingues PM, Amado FM (2006) Two-dimensional electrophoresis study of in vitro pellicle formation and dental caries susceptibility. Eur J Oral Sci 114(2):147–153

143. Vitorino R, Calheiros-Lobo MJ, Williams J, Ferrer-Correia AJ, Tomer KB, Duarte JA, Domingues PM, Amado FM (2007) Peptidomic analysis of human acquired enamel pellicle. Biomed Chromatogr 21(11):1107–1117

144. Vitorino R, Calheiros-Lobo MJ, Duarte JA, Domingues PM, Amado FM (2008) Peptide profile of human acquired enamel pellicle using MALDI tandem MS. J Sep Sci 31:523–537

145. Vitorino R, Barros A, Caseiro A, Domingues P, Duarte J, Amado F (2009) Towards defining the whole salivary peptidome. Prot Clin Appl 3:528–540. https://doi.org/10.1002/prca.200800183

146. Vitorino R, Guedes S, Manadas B, Ferreira R, Amado F (2012) Toward a standardized saliva proteome analysis methodology. J Proteome 75(17):5140–5165. https://doi.org/10.1016/j.jprot.2012.05.045. Epub 2012 Jul 15

147. Vitorino R, Barros AS, Caseiro A, Ferreira R, Amado F (2012) Evaluation of different extraction procedures for salivary peptide analysis. Talanta 94:209–215. https://doi.org/10.1016/j.talanta.2012.03.023. Epub 2012 Mar 30

148. Vlassov AV, Magdaleno S, Setterquist R, Conrad R (2012) Exosomes: current knowledge of their composition, biological functions, and diagnostic and therapeutic potentials. Biochim Biophys Acta 1820(7):940–948. https://doi.org/10.1016/j.bbagen.2012.03.017. Epub 2012 Apr 1

149. Vukosavljevic D, Custodio W, Buzalaf MA, Hara AT, Siqueira WL (2014) Acquired pellicle as a modulator for dental erosion. Arch Oral Biol 59(6):631–638. https://doi.org/10.1016/j.archoralbio.2014.02.002. Epub 2014 Feb 10

150. Wainwright WW (1934) Human saliva II.A technical procedure for calcium analysis. J Dent Res 14:425–434

151. WHO/IARC (2007) Common Minimum Technical Standards and Protocols for Biological Resource Centres Dedicated to Cancer Research, WorkGroup Report 2, Caboux E, Plymoth A and Hainaut P, Editors, IARC.

152. Wolff A, Davis RL (1991) Universal collector for submandibular-sublingual saliva. U.S. Pat. N° 5.050.616

153. Wolff A, Begleiter A, Moskona D (1997) A novel system of human submandibular/sublingual saliva collection. J Dent Res 76(11):1782–1786

154. Wolff A, Joshi RK, Ekström J, Aframian D, Pedersen AM, Proctor G, Narayana N, Villa A, Sia YW, Aliko A, McGowan R, Kerr AR, Jensen SB, Vissink A, Dawes C (2017) A guide to medications inducing salivary gland dysfunction, Xerostomia, and subjective Sialorrhea: a systematic review sponsored by the world workshop on Oral medicine VI. Drugs R D 17(1):1–28. https://doi.org/10.1007/s40268-016-0153-9

155. Xiao H, Wong DT (2012) Proteomic analysis of microvesicles in human saliva by gel electrophoresis with liquid chromatography-mass spectrometry. Anal Chim Acta 723:61–67. https://doi.org/10.1016/j.aca.2012.02.018. Epub 2012 Feb 19. uvwxyz

156. Xiao X, Liu Y, Guo Z, Liu X, Sun H, Li Q, Sun W (2017) Comparative proteomic analysis of the influence of gender and acid stimulation on normal human saliva using LC/MS/MS. Proteomics Clin Appl 11(7–8). https://doi.org/10.1002/prca.201600142. Epub 2017 Mar 21

157. Yan W, Apweiler R, Balgley BM, Boontheung P, Bundy JL, Cargile BJ, Cole S, Fang X, Gonzalez-Begne M, Griffin TJ, Hagen F, Hu S, Wolinsky LE, Lee CS, Malamud D, Melvin

JE, Menon R, Mueller M, Qiao R, Rhodus NL, Sevinsky JR, States D, Stephenson JL, Than S, Yates JR, Yu W, Xie H, Xie Y, Omenn GS, Loo JA, Wong DT (2009) Systematic comparison of the human saliva and plasma proteomes. Proteomics Clin Appl 3(1):116–134

158. Yang J, Wei F, Schafer C, Wong DT (2014) Detection of tumor cell-specific mRNA and protein in exosome-like microvesicles from blood and saliva. PLoS One 9(11):e110641. https://doi.org/10.1371/journal.pone.0110641. eCollection 2014

159. Yao Y, Grogan J, Zehnder M, Lendenmann U, Nam B, Wu Z, Costello CE, Oppenheim FG (2001) Compositional analysis of human acquired enamel pellicle by mass spectrometry. Arch Oral Biol 46:293–303

160. Yao Y, Berg EA, Costello CE, Troxler RF, Oppenheim FG (2003) Identification of protein components in human acquired enamel pellicle and whole saliva using novel proteomics approaches. J Biol Chem 278:5300–5308

161. Zhang J, Zhong LJ, Wang Y, Liu LM, Cong X, Xiang RL, Wu LL, Yu GY, Zhang Y (2017) Proteomic analysis reveals an impaired Ca2+/AQP5 pathway in the submandibular gland in hypertension. Sci Rep 7(1):14524. https://doi.org/10.1038/s41598-017-15211-0

162. Zheng X, Chen F, Zhang J, Zhang Q, Lin J (2014) Exosome analysis: a promising biomarker system with special attention to saliva. J Membr Biol 247(11):1129–1136. https://doi.org/10.1007/s00232-014-9717-1. Epub 2014 Aug 19

163. Zheng X, Chen F, Zhang Q, Liu Y, You P, Sun S, Lin J, Chen N (2017) Salivary exosomal PSMA7: a promising biomarker of inflammatory bowel disease. Protein Cell 8(9):686–695. https://doi.org/10.1007/s13238-017-0413-7. Epub 2017 May 18

164. Zimmerman JN, Custodio W, Hatibovic-Kofman S, Lee YH, Xiao Y, Siqueira WL (2013) Proteome and peptidome of human acquired enamel pellicle on deciduous teeth. Int J Mol Sci 14(1):920–934. https://doi.org/10.3390/ijms14010920

165. Zlotogorski-Hurvitz A, Dayan D, Chaushu G, Korvala J, Salo T, Sormunen R, Vered M (2015) Human saliva-derived exosomes: comparing methods of isolation. J Histochem Cytochem 63(3):181–189. https://doi.org/10.1369/0022155414564219. Epub 2014 Dec 3

Chapter 3
MS-Based Proteomic Analysis of Serum and Plasma: Problem of High Abundant Components and Lights and Shadows of Albumin Removal

Monika Pietrowska, Agata Wlosowicz, Marta Gawin, and Piotr Widlak

Blood serum or plasma proteome is a gold mine of disease biomarkers. However, complexity and a huge dynamic range of their components, combined with multiple mechanisms of degradation and posttranslational modifications, further complicated by the presence of lipids, salts, and other metabolites, represent a real challenge for analytical sensitivity, resolution, and reproducibility. This problem exists particularly in the case of potential disease-specific markers, most typically represented by low-abundant proteins (LAPs), whose detection is usually impaired by the dominance of albumins, immunoglobulins, and other high-abundant serum/plasma proteins (HAPs). Hence, analysis of biomarker candidates in serum/plasma samples frequently requires separation of their components, usually including depletion of albumin in a fraction of interest. Such "preprocessing" of serum/plasma specimens is critical in proteomic analysis based on mass spectrometry. This approach is very potent; nevertheless a wide range of protein concentrations in serum/plasma represents a particular challenge, since high-abundant proteins (mostly albumin) dominate in a sample subjected to mass spectrometry and suppress peptide ions originating from low-abundant proteins, thus limiting probability and reliability of their detection. An emerging approach in serum-/plasma-based biomarker-oriented studies is the proteome component of exosomes – nanovesicles secreted by cells and involved in multiple aspects of intercellular communication. However, the presence of albumin, frequent contaminant of exosomes isolated from human serum/plasma, represents a real challenge also in this type of study. A similar problem is encountered in proteomic studies based on exosomes obtained in in vitro experiments where culture media are normally supplemented with fetal bovine serum containing growth factors and hormones. In this case exosomes are frequently

M. Pietrowska · A. Wlosowicz · M. Gawin · P. Widlak (✉)
Maria Sklodowska-Curie Institute - Oncology Center, Gliwice Branch, Gliwice, Poland
e-mail: piotr.widlak@io.gliwice.pl

© Springer Nature Switzerland AG 2019
J.-L. Capelo-Martínez (ed.), *Emerging Sample Treatments in Proteomics*, Advances in Experimental Medicine and Biology 1073, https://doi.org/10.1007/978-3-030-12298-0_3

contaminated with bovine serum albumin and other bovine serum proteins which should be removed before proteomic analysis of exosome cargo.

This paper addresses the issue of preparation of serum/plasma specimens for mass spectrometry-based proteomic analyses. Methods for separation and depletion of albumin (and other high-abundant proteins) from blood serum/plasma samples or exosome preparations are presented and critically reviewed.

This work was supported by the National Science Centre (Poland), grants 2013/11/B/NZ7/01512 and 2016/22/M/NZ5/00667.

3.1   Plasma and Serum: Liquid Fraction of Blood

Plasma is considered as the medium of blood, in which blood cells and other morphological components are suspended. Serum is similar in composition to plasma but excludes a majority of factors involved in blood coagulation (clotting). This includes fibrinogen, which is converted into fibrin (active form of fibrinogen) during coagulation. Different types of plasma/serum proteins have very diversified concentration. The most abundant plasma proteins (with their contribution to the total protein content) are albumin (about 60%), globulins (35%), fibrinogen (4%), lipoproteins (~1%), and iron binding/transferring proteins (~1%). In general, 22 proteins account for about 99% of all plasma/serum proteins. The remaining 1% of blood proteins is composed of a few hundreds (or thousands) of low abundance circulatory proteins as well as proteins released by live, apoptotic, and necrotic cells. Most of blood proteins are secreted by the liver and intestines except for the gamma globulins, synthesized by the immune system [1–3].

There is a constant debate on what type of "liquid fraction" of blood, i.e., plasma or serum, analytical studies should be performed for (Fig. 3.1). While serum is considered the gold standard in many types of applications, laboratories must consider the turn-around time, which is a metric for lab performance and, more importantly, plays a critical role in patient care. Serum is free of clotting proteins but contains the clotting metabolites that result from the clotting process. It is a cleaner sample typically free of cells and platelets, since they are trapped in the fibrin meshwork of the clot. Plasma, on the other hand, is the liquid portion of blood that has been prevented from clotting by different types of anticoagulants and reflects the blood better as it circulates in the body. Although it has the advantage over serum that testing is not delayed by waiting 30 minutes for a clot to be formed, it is typically contaminated by platelets and cellular elements that have a potential to alter analytical results. Important features of plasma and serum specimens are compared in Table 3.1 [4, 5].

In general, serum/plasma is a body fluid reflecting the physiological state of an organism, hence it is frequently used in research on biomarkers of disease processes. Owing to the stabilizing role of serum/plasma and maintaining the homeostasis of an organism, the observed changes in composition are rather small and are of rather quantitative than qualitative character. Obviously, analysis of proteome of a target

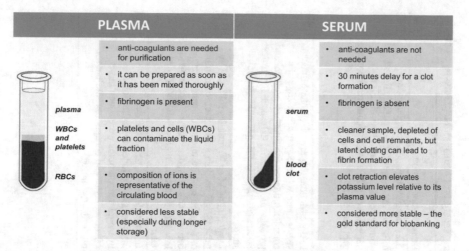

PLASMA		SERUM	
	• anti-coagulants are needed for purification		• anti-coagulants are not needed
	• it can be prepared as soon as it has been mixed thoroughly		• 30 minutes delay for a clot formation
plasma	• fibrinogen is present	*serum*	• fibrinogen is absent
WBCs and platelets	• platelets and cells (WBCs) can contaminate the liquid fraction		• cleaner sample, depleted of cells and cell remnants, but latent clotting can lead to fibrin formation
RBCs	• composition of ions is representative of the circulating blood	*blood clot*	• clot retraction elevates potassium level relative to its plasma value
	• considered less stable (especially during longer storage)		• considered more stable – the gold standard for biobanking

Fig. 3.1 Comparison of serum and plasma as material used in the diagnostics

tissue is a more reliable source of information about pathology than analysis of a substitute tissue (i.e., serum/plasma). Nevertheless, in the case of screening tests, early disease detection in preclinical stage, detection of micrometastases, or diagnostics of "nonsurgical" cancers, analysis of blood proteome is an indispensable diagnostic method.

3.2   Blood-Derived Exosomes

In recent years scientific research has focused not only on plasma and/or serum as such but also on new potential sources of biomarkers derived from this material. An important source of disease markers could be vesicles appearing de novo in serum/plasma as a response to a disease process, which (among others) include exosomes. Exosomes have emerged as a novel class of intercellular signal mediators that are involved in various pathological conditions. They are a kind of cell-released membrane-enclosed nanovesicles, which can be released by all types of cells including cancer cells, fibroblasts, immune cells, and mesenchymal cells. The content of exosomes, including a cargo of different nucleic acids, lipids, and proteins, is mostly defined by their parent cells. Exosomes were initially thought to be a mechanism for removing unneeded proteins. However, recent studies have revealed that they are actually specialized in long-distance intercellular communication facilitating a transfer of proteins, functional mRNAs, and microRNAs for subsequent protein expression in target cells. Being able to transfer their contents in a stable way to distant destinations, exosomes were proved to be an effective mode of a cross talk among cells far apart and to be involved in multiple physiological and pathological processes. Numerous experiments demonstrated

Table 3.1 Examples of depletion/enrichment strategies employed in sample pre-treatment prior to MS-based analysis of human serum/plasma proteome

Material	Depletion/enrichment strategy	Chromatographic separation	MS system	Remarks/conclusions	Ref.
Serum/plasma	1. Precipitation of HAPs with ACN/0.1% TFA 2. Proteomics-30® resin column material: restricted access materials (RAM) combined with immobilized metal affinity chromatography (IMAC) 3. ProteoMiner® peptide ligand affinity beads for equalization	Weak cationic exchange (CM10) on ProteinChip arrays	SELDI-TOF	All strategies provided complementary information; however only peptide ligand affinity beads combined both efficient depletion of HAPs and enrichment in protein/peptide peaks	[22]
Plasma	1. Single enrichment: ProteoMiner® 2. Single depletion: ProteoPrep20® 3. Multistep depletion: ProteoExtract® + ProteoPrep20®	SCX fractionation (5 fractions) followed by reversed-phase ChipCube nanoLC	Q-TOF	LAPs enrichment approach (ProteoMiner®) seems more suitable as the first stage of a complex multistep fractionation protocol and provides more material to be used for further fractionations; it is cheaper and technically simpler	[17]
Serum	1. MARS6 immunodepletion LC column 2. MARS14 immunodepletion LC column 3. ProteoPrep20® immunodepletion LC column	RP nanoLC	LTQ-Orbitrap Velos	Depletion of 14 or 20 proteins followed by LC-MS/MS resulted in good reproducibility of detected proteins and significant overlap between columns; detection of moderate to lower abundance proteins (sub-ng/mL–µg/mL)	[50]
Serum	1. Size exclusion-affinity hydrogel nanoparticles 2. Vivaspin® centrifugal concentrators based on ultrafiltration membranes 3. ProteoMiner® peptide ligand affinity beads for equalization	RP nanoLC	Orbitrap LTQ-XL	Immunodepletion protocols were intentionally excluded from the study; hydrogel nanoparticles were found more effective, faster and cheaper option than the other two technologies	[16]

Plasma	1. single immunodepletion: ProteoPrep20® 2. tandem immunodepletion: ProteoPrep20®	RP nanoLC	LTQ-Orbitrap Elite	Proteins at concentration range of ng/mL were identified; application of longer RP-LC gradients and columns offers good compromise between effort, throughput and reproducibility	[51]
Serum/plasma	1. precipitation of albumin with a mixture of 0.1% trichloroacetic acid in isopropanol (IPA-TCA) 2. Pierce albumin depletion kit (dye-protein interaction)	RP UHPLC	QQQ operated in SRM mode	The proposed IPA-TCA method is dedicated to medium-abundant proteins (LOQ ca. 50 ng/mL)	[23]
Plasma	1. single enrichment: ProteoMiner® 2. tandem depletion: Seppro IgY14 3. single depletion: Seppro IgY14	High-pH RP LC or SDS-PAGE	Q-Exactive	Coupling single IgY14 immunodepletion and high-pH RPLC with LC-MS/MS yielded the greatest depth of analysis; achievable concentration range: from sub-ng/mL to µg/mL	[52]

Abbreviations: ACN acetonitrile, *HAPs* high-abundant proteins, *IPA* isopropanol, *LAPs* low-abundant proteins, *LOQ* limit of quantification, *MARS* multiple affinity removal system, *QQQ* triple quadrupole, *Q-TOF* quadrupole-time-of-flight tandem mass spectrometer, *RP* reversed-phase, *SELDI* surface-enhanced laser desorption/ionization, *SCX* strong cation exchange, *SRM* selective reaction monitoring, *TCA* trichloroacetic acid, *TFA* trifluoroacetic acid that tumor-related exosomes (TEXs) can induce immune surveillance in the micro-environment in vivo and in vitro. They can interfere with maturation of DC cells, impair NK cell activation, induce myeloid-derived suppressor cells, and educate macrophages into a pro-tumor phenotype. A part of exosomal cargo simply reflects the composition of a parent cell, while another set of cellular components is selectively enriched, including endosome-associated proteins (e.g., Alix), proteins associated with lipid raft (flotillin and glycosylphosphatidylinositol-anchored protein), and tetraspanins (e.g., CD63, CD9, CD81, and CD82). However, one should keep in mind that although these markers play a crucial role in identification and purification of exosomes, they are not unique to these vesicles. A review of methods of isolation and purification of exosomes can be found in numerous papers [6–11]; however not all of them take into consideration compatibility with mass spectrometry techniques. One of the main challenges in analysis of isolated exosomes is co-isolated high-abundant proteins (HAPs). This is why much attention is paid to preparation of such difficult samples for mass spectrometry analysis.

3.3 Mass Spectrometry-Based Analysis of Serum or Plasma Specimens

Mass spectrometry (MS) techniques have evolved into an indispensable tool in proteome research owing to their high throughput, selectivity, sensitivity, and versatility. However, in the case of plasma or serum specimens characterized with high complexity and extremely wide range of concentrations of individual proteins/peptides present in a sample (the order of magnitude above 10^{10}) [1], even the most technologically advanced systems cannot ensure detection of lower abundant species. This extremely wide range of protein abundance is the major limiting factor in the overall result of proteomic analysis of serum/plasma with the use of MS. Regardless of the applied ionization technique, high-abundant proteins steal the show in MS-based identification due to strong ion suppression caused by the presence of peptides originating from dominating proteins (e.g., albumin, immunoglobulins, apolipoproteins). This phenomenon results in loss of information of low-abundant proteins/peptides (e.g., interleukins, cytokines, other signal peptides), which are usually the main subject of interest of a research. Therefore, workflows dedicated to MS analysis of serum/plasma proteome include various sample pre-treatment steps aimed at depletion of highly abundant proteins and/or enrichment of low-abundant proteins of interest, and/or fractionation of peptides obtained after proteolytic digestion [12]. A general workflow in MS-based proteomic analysis of serum/plasm has been schematically presented in Fig. 3.2. This subsection deals with different approaches that have been used to solve this problem. More comprehensive information about commercially available methods of albumin removal is given below in Sect. 3.4.

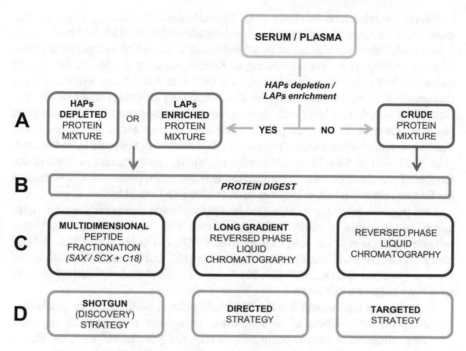

Fig. 3.2 General workflow in MS-based proteomic analysis of serum/plasma: (A) sample pre-treatment; (B) proteolytic digestion; (C) peptide fractionation; (D) mass spectrometry analysis; HAPs, high-abundant proteins; LAPs, low-abundant proteins; SAX, strong anion exchanger; SCX, strong cation exchanger; C18, octadecyl carbon chain-bonded silica

The above-mentioned problem of masking effect of high-abundant proteins is an issue generally well-recognized by mass spectrometrists. However, although much effort has been made in serum/plasma proteome research to find most effective and efficient strategy of HAPs depletion and/or LAPs enrichment, so far no consensus has been reached on a universal sample pre-treatment workflow. Comparison of various available methods in terms of efficiency of HAPs removal from serum/plasma samples and increasing accessibility of LAPs to MS detection/quantification was performed by several research groups. Compared strategies were ranked on the basis of different criteria, including efficiency of HAPs removal, the number of detected/identified proteins, as well as their cost and required workload.

In general, immunodepletion techniques with the use of liquid chromatography columns or spin columns have been so far the most commonly used tool for HAPs removal from serum/plasma samples prior to MS-based analysis. There are several commercially available solutions offered by different vendors, e.g., ProteoPrep20® LC column and spin columns (Sigma-Aldrich); Seppro® IgY14 LC columns, spin columns, and 96-well plates (Sigma); and MARS® LC columns (Agilent). However, one must bear in mind potential losses of low-molecular weight proteins owing to their interactions with larger proteins acting as carriers and to unspecific binding of LAPs to antibodies [13–15]. Therefore, some researchers intentionally exclude immunodepletion-based strategies [16]. ProteoMiner® technology (BioRad) has gained much interest as an alternative to immunodepletion in MS-based proteomics. It is an enrichment strategy based on a combinatorial library of hexapeptides bound to a chromatographic support. Owing to limited capacity of the sorbent, HAPs quickly saturate dedicated ligands, and their excess is washed out, while LAPs get concentrated; therefore this strategy is also referred to as protein equalization. It is technically simple and less costly than immunodepletion-based LC columns [17]; however, Zhao et al. reported poor enrichment capacity for acidic proteins and macromolecular proteins when compared to depletion with Seppro® IgY14 column [18]. Hakimi et al. found ProteoMiner® enrichment more efficient in terms of the number of identified proteins in comparison with ProteoPrep® column, nevertheless the former solution required larger sample volumes (>1 mL) [19].

Less popular solutions to reduction of serum/plasma complexity include inter alia (i) precipitation of HAPs with cold acetone [20], acetonitrile [21, 22], or a mixture of trichloroacetic acid in isopropanol [23], (ii) application of centrifugal concentrators based on ultrafiltration membranes [16], and (iii) utilization of various types of modified solid supports [16, 22]. Selected examples of reported approaches have been collected in Table 3.1.

Regardless of the fact whether HAPs depletion/LAPs enrichment was performed or not, proteolytic digest of serum/plasma comprises thousands of peptides; therefore reduction of sample complexity is required prior to MS analysis. This is routinely performed by coupling a mass spectrometer with a liquid chromatograph equipped with C18 column. Reversed-phase liquid chromatography (RP-LC) provides not only peptide separation but also purification from salts and other potential interferents. Nano-flow LC systems are specially dedicated to peptide separations, and typical capillary columns are 10–15 cm long. Further reduction in sample complexity can be obtained by preceding RP-LC separation with peptide fractionation with the use of ion exchange chromatography. Owing to amphoteric properties of peptides, both strong anion exchangers (SAX) and strong cation exchangers (SCX) can be employed.

Although multidimensional pre-fractionation of peptides has a great potential to maximize the depth of proteomic analysis, it may be omitted in the case of long-gradient liquid chromatography separations. Shi et al. proposed a method for quantitative analysis of human serum proteins at concentrations ranging from pg/mL to ng/mL with the use of long-gradient selected reaction monitoring (LG-SRM) mass spectrometry [24]. Application of a 150-cm-long liquid chromatography capillary column and gradient time of 300 min enabled SRM-based quantification of eight endogenous proteins in tryptic digest of serum samples without the need of previous depletion of high-abundant proteins. Further enhancement in sensitivity of the proposed LG-SRM method was achieved by incorporation of IgY14 immunoaffinity depletion to the workflow (tantamount to removal of 14 most abundant proteins), which resulted in reduction of interferences from serum matrix and ca. fourfold increase in S/N ratio. This gives potential for quantification of proteins present in human serum/plasma at low- to sub-ng/mL level, yet increases the risk of losing some LAPs.

In parallel to tailored sample pre-treatment, the depth of proteomic analysis of such complex biological samples as serum and plasma is strongly influenced by application of a suitable MS strategy [25–27]. Shotgun strategy has been widely used in analysis of serum/plasma proteome, especially in the context of biomarker research [27]. This strategy does not require any prior knowledge of composition of a sample; hence it is especially suited for discovery experiments, not only protein identification but also analysis of their posttranslational modifications. Application of stable isotope labeling methods of quantification or a label-free approach enables comparison between two or more biological states in terms of relative protein abundances. Typically, a protein mixture is subjected to proteolytic digestion, and the resulting digest is subsequently analyzed with the use of an LC-MS/MS system. Peptides, separated in a chromatographic column, enter a mass spectrometer which is operated in data-dependent acquisition (DDA) mode (also known as information-dependent acquisition (IDA)): a spectrometer continuously registers a full mass spectrum at a given time point of LC elution (i.e., a survey scan) and automatically selects a list of precursor ions (intact peptides) which are subjected to controlled fragmentation. The selection is performed basing on information derived from the MS spectrum – typically, the most intense peaks are selected. As a result, characteristic product-ion spectra are acquired which are subsequently used for peptide identification. Robust, high-resolution hybrid mass spectrometers like linear ion trap-Orbitrap, quadrupole-Orbitrap, and quadrupole-time-of-flight (Q-TOF) spectrometers are most frequently employed. It is the most informative strategy in terms of qualitative data and analysis of relative abundance of proteins in different biological states; however, it suffers from high costs of measurement and computation time.

In directed MS strategy protein identification and quantification are realized as separate processes in two experiments: a list of proteins identified in the first process serves for construction of a list of targets (so called inclusion list) for the second process. Consequently, such a workflow requires at least two LC-MS or LC-MS/MS measurement for each sample. Nevertheless, the possibility of excluding from analysis, e.g., unwanted most intense signals originating from contaminants, enables deeper proteome exploration and focusing on less abundant species.

Targeted MS is of special significance in clinical tests where accurate, reproducible, and sensitive quantitative measurements of protein markers are required [28]. Analyses are typically realized with the use of a triple quadrupole MS systems where the first quadrupole (Q1) serves as a mass filter for preselected set of peptides characteristic for proteins of interest, which once detected undergo a controlled and specific fragmentation in the second quadrupole. The third quadrupole (Q3) is set to monitor pre-defined fragments of a given peptide. Each "precursor ion-fragment ion" pair is called a transition. In this way, during one work cycle of a mass spectrometer (depending on the width of a chromatographic peak) dozens of transitions can be monitored in so-called multiple reaction monitoring (MRM), also known as selected reaction monitoring (SRM) mode. Targeted MS exceeds the remaining strategies in terms of achieved selectivity, sensitivity, reproducibility, and quantitation accuracy; however its effectiveness is lower.

Along with the birth and development of SWATH technology (Sequential Windowed Acquisition of All Theoretical Fragment Ion Mass Spectra) [29], mass spectrometry-based proteomics has gained an exceptionally powerful tool which has a good chance of revolutionizing in-depth analysis of complex biological systems, especially in clinical applications [30, 31]. Being a data-independent acquisition (DIA) approach, it overcomes the problem of stochastic selection of peptide precursors encountered in typical data-dependent approaches characteristic for shotgun experiments. Thanks to utilization of a wider precursor ion selection window in quadrupole Q1 (in comparison with MRM measurements, e.g., 25 Da), more ions enter the collision chamber; hence the resulting MS/MS spectrum, being a compilation of fragment spectra of all analytes within this m/z window, is much more complicated and informative. Upon shifting Q1 window within a selected m/z region, a full MS/MS spectrum is collected with LC-compatible cycle time. Since fragment ions are registered with high resolution, high-quality extracted ion chromatograms (XIC) are created once data acquisition is finished and MRM-like data is obtained which can be subsequently integrated for quantitative analysis. Consequently, SWATH MS combines high throughput of shotgun approaches with excellent reproducibility, sensitivity, and accuracy of MRM-based measurements.

High effectiveness of SWATH MS in protein discovery and its targeted character are highly beneficial also in analysis of serum/plasma proteins. Liu et al. implemented this technology in a study on quantitative variability of plasma proteins in a human twin population [32]. SWATH MS measurements were performed for peptide mixtures obtained via tryptic digestion of crude plasma samples. No prior peptide fractionation was performed to minimize confounding factors introduced during sample handling. The developed SWATH MS method enabled quantification of 1904 peptides defining 342 unique plasma proteins in 232 plasma samples within protein concentration range of 10^6 and the lowest concentration level of several ng/mL.

The group of R. Aebersold applied SWATH MS also to quantitative measurements of N-linked glycoproteins in human plasma [33]. The developed method was confronted with an SRM-based workflow in terms of limit of quantification, reproducibility, and quantification accuracy. Plasma samples were subjected to solid-phase extraction of N-linked glycopeptides (SPEG) with the use of Affi-gel hydrazine resin (Bio-Rad): retained glycoproteins were digested with trypsin directly on SPE resin, and N-linked glycopeptides were released from the bed with PNGase F. The limit of quantification reached at peptide level with the use of the developed SWATH MS method was only three times lower than the one obtained in SRM measurements, giving a concentration of 5–10 ng protein/mL in plasma, and the obtained reproducibility was comparable. Nevertheless, in comparison with SRM approach SWATH MS offered greater opportunities in terms of the number of quantified peptides in a single analysis. Without a proper enrichment strategy, quantification of low-abundant protein glycoforms is problematic in such a complex system as body fluids. This challenge was addressed by Sanda and Goldman who utilized SWATH MS-based data-independent analysis of IgG glycoforms in unfractionated human plasma [34]. Tryptic peptides of plasma proteins were obtained via in-solution digestion and LC-MS/MS analyzed without any further processing. The proposed method is based on utilization of Y-ions generated during fragmentation of glycopeptides using collision-induced dissociation (CID) process; however the employed collision energy was lower than in typical peptide fragmentation. Such soft fragmentation conditions favored high yield of glycopeptide-derived Y-ions (up to 60% of precursor ion intensity) and minimized interferences resulting from further fragmentation of N-glycans and from fragmentation of non-glycosylated peptides. The developed method allowed selective quantification of multiple IgG glycoforms in unfractionated human plasma.

3.4 Commercial Tools for Albumin Removal

Attempts at specific removal of albumin from plasma/serum samples are of prime importance, since the presence of albumin can hinder identification of other proteins not only in MS-based approaches but also in 2-D electrophoresis and Western blotting. Several commercially available kits were described in literature as a tool for albumin removal from biological fluid samples. Throughout the years of struggling with albumin "contamination" many techniques were investigated, nevertheless most of commercial tools are based on high affinity of albumin for different materials. Numerous published studies include sample purification with dye affinity resins or immunoaffinity columns [15, 35, 36]. Some of the most common modified resins have immobilized textile dye ligands, like Cibacron Blue F3GA. This anionic anthraquinone dye is covalently bound to agarose beads allowing depletion of albumin. It is commercially used in several separation systems: Pierce™ Albumin Depletion Kit® (Thermo Fisher Scientific), Montage™ Albumin Deplete Kit® (Millipore) or Aurum™ Affi-Gel® Blue Columns (Bio-Rad). Dye ligand column kits were found effective in separating human serum albumin from several biofluid samples.

Immunoaffinity capturing of proteins is an alternative method of sample purification. It is based on binding of certain proteins to monoclonal or polyclonal antibodies linked with resin beads and is characterized with higher specificity compared to dye-based resins [37]. On the other hand, immunoaffinity techniques may provide lower capacity than dye resins [2], whereas this parameter is very important for efficient albumin binding and minimizing the carryover effect. Separation of serum analyzed for the presence of SCCA1 (squamous cell carcinoma antigen 1) revealed that HiTrap Blue/HiTrap Protein G columns (Amersham), Albumin Removal Column (Merck), and Aurum™ Serum Protein (Bio-Rad) provided good overall albumin depletion and higher loading capacity in comparison with immunoaffinity anti-HAS/POROS Protein G column (Applied Biosystems) [38]. Some ready-to-use kits use two different types of beads combined together. Aurum™ Serum Protein Mini Kit (Bio-Rad) and ProteoPrep® Albumin IgG Depletion Kit (Sigma) offer immunoaffinity capture columns for simultaneous depletion of HSA and IgG. Commercially available affinity kits are constantly being improved in order to enhance the number of eliminated main serum/plasma proteins. Polaskova et al. performed a comparative study to evaluate improvement in reduction of complexity of plasma samples with the use of six commercial depletion products [35]. ProteoPrep® 20 Plasma Immunodepletion Kit (Sigma) showed good resolution improvement in 2-D electrophoresis of plasma samples. Twelve out of 20 main proteins claimed by the manufacturer could be removed; however traces of apolipoprotein A-I, ceruloplasmin, and IgG were still present in the flow-through fractions. ProteomeLab™ IgY-12 (Beckman Coulter) provided relatively better depletion than the aforementioned kit – 10 out of 12 claimed HAPs. This kit uses polyclonal IgY antibody as a ligand which binds to multiple epitopes; however it is an avian antibody, and therefore it is not specific for several mammalian proteins, such as complement factors, rheumatoid factor, IgM, Fc receptor, and protein A or G.

Despite the fact that commercial kits enable effective binding of albumin, and that columns are reusable, there are some limitations which do not allow to use them for every biofluid analysis. First of all, co-binding of some low-abundant proteins to resins was reported. For example, resins with Cibacron Blue dye ligand remove nucleotide-binding proteins as well as NAD, FAD, and ATP along with albumin [38, 39]. Lack of specificity is not the only obstacle in this field. It has to be noted that the main role of albumin is to transport active proteins in the circulatory system. It has natural affinity to bind many proteins of interest, such as enzymes, hormones, etc. Even if a resin has high specificity mainly towards albumin (even >80% of albumin is removed as the manufacturers declare), the problem lies in albumin itself. As a result, the loss of desirable proteins may significantly lower sensitivity of an analysis. Liu et al. showed high albumin depletion (98%) from serum/plasma samples with markedly high loss of target proteins (over 50%) [23].

3.5   Removal of Albumin from Exosome Preparation

The interest in developing an ideal method for albumin depletion from serum or plasma is indeed of high importance. This issue becomes especially challenging in terms of analysis of proteome of vesicles present in blood, to which growing attention has been given recently. For the last decade researchers have been trying to develop a method which would allow to obtain morphologically intact and biologically active low-abundant vesicle preparations for high-throughput proteomic analysis free from HAPs contamination. As exosomes are considered potential biomarkers for cancer and could be used in future as a "liquid biopsy" for screening patients and predicting clinical outcome, it is also crucial to find an optimal method of their purification and analysis. However, to the best of our knowledge there is still a lack of a standardized method of exosome purification from serum or plasma specimens. Nevertheless, there are miscellaneous techniques available which were widely tested on serum or plasma-derived exosomes and some of them deserve particular attention [6, 40, 41]. Interferences from albumin and other HAPs present in serum/plasma samples and their consequences in mass spectrometry analysis of plasma/serum is a well-recognized problem. However, it seems that too little interest is paid to this issue in MS-based analysis of exosomal protein cargo as well as in molecular biology techniques for exosome characterization. This challenging problem reflects not only analysis of exosomes purified from biofluids but also exosomes purified from culture media. In fact, it is frequently underappreciated that fetal bovine serum, a standard supplement of cell culture media containing growth factors and hormones, can generate serious problems in MS-based analysis of exosome proteome. Hence, in addition to methods described above in Sects. 3.3 and 3.4, others have been developed recently for albumin removal from serum/plasma prior to exosomes protein identification.

Currently, the applied techniques for exosome purification from biofluids or cell culture medium involve ultracentrifugation, precipitation, density gradient separation, immunoaffinity, or size-exclusion chromatography (SEC). At present, the most commonly used methods for exosome isolation include ultracentrifugation alone [40] and combined with sucrose gradient, as well as immunomagnetic bead isolation (e.g., magnetic activated cell sorting (MACS)) [41]. Moreover, there are many commercial kits available for extraction of exosomes. ExoQuick™ (System Bioscience) or Exo-spin™ systems (Cell Guidance Systems) offer exosome purification based on a reaction between exosome nanoparticles and a precipitating buffer with subsequent resuspension of the obtained exosome pellet. Commercially available size-exclusion columns (qEV™, Izon Science Ltd.) use cross-linked beads which retain small particles and let bigger particles to flow through the medium faster. According to these assumptions, exosomes elute before albumin (66 kDa) and IgG fragments, which allows to isolate exosomes derived from cell culture media and from biofluids. The use of magnetic microbeads with attached antibodies becomes a superior technique, since antibodies are conjugated precisely with one specific antigen, such as membrane proteins present on exosomes. Manufacturers offer ready-to-use magnetic beads coupled with certain ligands, like epithelial cell adhesion molecule EpCAM (CD326) suitable for culture media and biofluids (MACS®, Miltenyi Biotec) [7, 42]. Examples of different approaches to exosome purification are summarized in Table 3.2.

Unfortunately, the most popular methods of exosome isolation do not pay attention to the problem of co-isolation of abundant serum/plasma proteins with exosomes while subjecting the samples to further proteomic assays. For example, ExoQuick™ and Exo-spin™ protocols were tested against qEV™ for plasma exosomes isolation, and precipitation of exosomes derived from human plasma resulted in significantly higher yields of particles with size of <100 nm, while compared to SEC columns. However, precipitation-based techniques were shown to suffer from contamination with co-precipitating proteins and a high degree of albumin content was present only in samples isolated with ExoQuick™and Exo-spin™ [6] (Table 3.2).This is a particular problem in the case of mass spectrometry-based proteomic studies of exosomes, where removal of overabundant sample components (e.g., albumin) is critical due to the problem of ion suppression.

Table 3.2 Examples of sample pre-treatment protocols employed prior to further analysis of proteome component of exosomes isolated from serum/plasma

Material	HAPs depletion/ enrichment strategy	Remarks	Ref.
Serum	Ultracentrifugation (5 cycles)	1. 1559 proteins quantified; 2. 8 proteins showed global treatment-specific changes in all of patients (1B51, vimentin, OBSL1, MARE2, LYSC, PLF4, CD71, PARVB)	[53]
Plasma	Not included in the study – exosomes isolated from plasma with ExoQuick™ kit	The majority of the ten most abundant identified proteins are in fact high-rank plasma proteins, such as haptoglobin, transferrin, apolipoproteins, ceruloplasmins, complement proteins	[46]
Serum	Isolation with use of ExoQuick™ kit followed by centrifugation (1500 × g, 10 min, 4 °C). The exosome pellet was then resuspended in 10 mM PBS in four times the volume of serum.	MALDI-TOF/TOF analysis revealed 52 proteins found in serum of patients with Kawasaki disease which are mostly high-abundant proteins	[48]
Plasma	Isolation of exosomes with UC protocol: plasma samples pre-filtered (0.22 μm) and centrifuged (120,000 × g) at different temperature and time conditions. Pellets were washed with PBS and centrifuged again (120,000 × g, 4 °C, 1 h).	1. All obtained exosome preparations were contaminated with albumin 2. UC performed for 1 h and 3 h resulted in isolation of intact vesicles (TEM analysis) while amorphous particles were predominant in samples after longer UC (6 h, 14 h) 3. Longer UC resulted in higher CD63 level but albumin content was also elevated. 4. Repeated UC step did not alter considerably the amount of exosomes in comparison with UC performed once	[54]
Plasma	Isolation of exosomes with SEC: plasma sample (diluted twofold with PBS) was loaded onto gravity-eluted columns of different matrices (Sepharose 2B, Sepharose CL-4B, or Sephacryl S-400). 1 mL fractions were eluted with PBS	Western blot analysis showed that albumin was present starting from fractions 6 eluted on Sepharose 2B and Sepharose CL-4B overlapping exosomes markers (TSG101 and CD63). Sephacryl S-400 retained albumin until fraction 7 was eluted, but in fact, albumin co-elution occurred despite used column matrix	[54]
Plasma	ExoQuick™, Exo-spin™ and qEV™ columns for depletion of albumin and other main blood components	1. qEV ™ provides lowest exosomes recovery but the highest purity from all tested kits 2. Between two precipitation techniques, Exo-spin™ gives less plasma contamination proteins than ExoQuick™	[6]

(continued)

Table 3.2 (continued)

Material	HAPs depletion/ enrichment strategy	Remarks	Ref.
Biofluid/ culture supernatant	Differential ultracentrifugation (300 × g, 10 min; followed by 2000 × g, 10 min; followed by 10,000 × g, 30 min; and at last 100,000 × g, 70 min – 2 cycles). Temp. 4 °C	Authors present complete protocol for the most common method of purified exosomes isolation. No comments are made in terms of reproducibility and yield of this method. However, based on our knowledge, this method is burdened with loss of exosomal material (results not present)	[40]
Serum	MACS procedure, using EpCAM, was applied for isolation of ovarian tumor-derived exosomes	EpCAM-positive exosomes could be isolated from sera of patients at different disease stages, and differences in exosomes concentration reflected clinical data on the cancer degree of advancement	[42]

Abbreviations: HAPs highly abundant proteins, *MACS* modified magnetic activated cell sorting, *SEC* size-exclusion chromatography, *PBS* phosphate buffer saline, *EpCAM* epithelial cell adhesion molecule, *TEM* transmission electron microscopy, *CD63* tetraspanin expressed on the surface of exosomes

Possible contamination of exosome fraction can occur due to co-precipitation of proteins, like albumin, as well as lipoproteins and lipid droplets [43]. All these molecules generally exist in serum along with exosomes and one should keep in mind that their presence may result in fallacious conclusions regarding identification of certain proteins and lipids as exosomal cargo. Next, serum contains not only exosome-size vesicles but also larger micro-vesicles or apoptotic bodies. It is possible that they can be co-isolated with exosomes during precipitation, which can mislead to the belief that identified proteins originated from exosomes but, in fact, they are derived from non-exosomal vesicles. It was also reported that exosomes themselves form aggregates in blood [44, 45], which results in loss of exosome material and poor recovery.

Another issue that has to be highlighted is that the obtained protein identification results are not always critically analyzed. Simple isolation of exosomes by precipitation with a polymer-based ExoQuick™ kit reported by Turay and co-workers [46] apparently resulted in heavily contaminated exosomes, as the majority of the identified highest abundance proteins, i.e. apolipoproteins, macroglobulins, albumin precursors, haptoglobin, ceruloplasmins, transferrin, and fibronectin, are in fact high-rank plasma proteins with concentrations ranging from $1 \cdot 10^3$ to dozens of mg/mL [47]. The authors did not comment on this result. The same approach was employed by Zhang et al. in their study on coronary artery dilatation caused by Kawasaki disease based on proteomic analysis of serum exosomes [48]. These and many other mistaken protein identifications of exosomal cargo augment the database of ExoCarta – there is a whole collection of high-abundant blood proteins which appear as exosomal proteins. This may be due to the attachment of albumin molecule to exosomes which can lead to wrong interpretation in the functional enrichment analyses of proteins performed with the use of some open-source solutions, e.g., FunRich based on ExoCarta database. However, some authors indicate HAPs, such as albumin, complement proteins or prothrombin as a particular group of exosomes-associated proteins which come from serum but associate extracellularly with exosomes after their secretion [49].

As it is shown in this section, each method has its pros and cons, and depending on the subject of interest, it is crucial to optimize the method of purification not only by choosing the right commercial kit but also by adjusting all parameters which affect efficient preparation especially while analyzing nanovesicles.

3.6 Summary

Commercially available methods of removal of albumin and other high-abundant proteins share one serious drawback, namely, they remove proteins/peptides transported by HAPs from analyzed material [39]. As a consequence, a group of blood proteins/peptides for whom albumin is a carrier protein is lost as a result of albumin depletion. Two alternative approaches are reported in papers where low molecular weight fraction of serum proteome was analyzed: whole serum is analyzed, or the analysis is preceded with albumin removal (and other high molecular mass proteins). A mass spectrometer registers all kinds of molecules present in body fluids. In the case of serum, the main detected proteins are of course albumin and immunoglobulins. Undoubtedly, the presence of albumin (which constitutes ca. 60% of all serum proteins) is a factor influencing the quality of analysis of proteins and peptides which are the actual subject of a study, and its removal from material prior to analysis is a considerable methodological problem. Moreover, the emerging possibility of isolation of circulating cancer cells or vesicles produced by normal or cancer cells from serum or plasma challenges scientists to combine their methods of isolation with subsequent purification.

MS techniques are known to be very demanding in terms of preparation of protein samples. Apart from new technologically available targets for analysis, such as exosomes, also purification strategies have changed. An increase in requirements for sample quality in MS analysis has been observed owing to constant technological advances in sensitivity of mass spectrometers and the resulting decrease in resistance to contamination and composition of utilized buffers. Currently, high-resolution mass spectrometry techniques require optimization of an analytical method prior to carrying out new analyses. Therefore, results obtained for the same samples will often be incomparable due to methodological differences. Mass spectrometry as a high-throughput technique provides with large datasets which are analyzed automatically with the use of a dedicated software. Statistical reliability parameters set up in a software is another bane of such analyses and inter-laboratory comparisons. Unfortunately, no golden standard has been developed in this matter so far.

Large diversity of methods for removal of albumin and other high-abundant serum/plasma proteins and their insufficient compatibility with MS techniques enforces the need for constant development and standardization of sample preparation methods and workflows for MS-based proteomic analysis of serum/plasma. In the case of exosome isolation, albumin removal is crucial not only for sample preparation for MS analysis but also for other techniques of molecular biology, such as fluorescence activated cell sorting (FACS) and microscopy. Albumin removal from biological samples is a tedious and difficult task; however in our opinion it is vital to conduct reliable and valuable scientific research. The most frequent choice of a quantitative assessment in the literature remains the Western blot technique, whereas it is realistic only in the case of analysis of a dozen of proteins or so, hence the attempts of utilization of mass spectrometry techniques.

References

1. Gianazza E, Miller I, Palazzolo L, Parravicini C, Eberini I (2016) With or without you - proteomics with or without major plasma/serum proteins. J Proteome 140:62–80. https://doi.org/10.1016/j.jprot.2016.04.002
2. Steel LF, Trotter MG, Nakajima PB, Mattu TS, Gonye G, Block T (2003) Efficient and specific removal of albumin from human serum samples. Mol Cell Proteomics 2:262–270. https://doi.org/10.1074/mcp.M300026-MCP200
3. Georgiou HM, Rice GE, Baker MS (2001) Proteomic analysis of human plasma: failure of centrifugal ultrafiltration to remove albumin and other high molecular weight proteins. Proteomics 1:1503–1506. https://doi.org/10.1002/1615-9861(200111)1:12<1503::AID-PROT1503>3.0.CO;2-M
4. Yu Z, Kastenmüller G, He Y, Belcredi P, Möller G, Prehn C, Mendes J, Wahl S, Roemisch-Margl W, Ceglarek U, Polonikov A, Dahmen N, Prokisch H, Xie L, Li Y, Wichmann HE, Peters A, Kronenberg F, Suhre K, Adamski J, Illig T, Wang-Sattler R (2011) Differences between human plasma and serum metabolite profiles. PLoS One 6:e21230. https://doi.org/10.1371/journal.pone.0021230
5. Jackson DH, Banks RE (2010) Banking of clinical samples for proteomic biomarker studies: a consideration of logistical issues with a focus on pre-analytical variation. Prot Clin Appl 4:250–270. https://doi.org/10.1002/prca.200900220
6. Lobb RJ, Becker M, Wen SW, Wong CS, Wiegmans AP, Leimgruber A, Möller A (2015) Optimized exosome isolation protocol for cell culture supernatant and human plasma. J Extracell Vesicles 4:27031. https://doi.org/10.3402/jev.v4.27031
7. Tauro BJ, Greening DW, Mathias RA, Ji H, Mathivanan S, Scott AM, Simpson RJ (2012) Comparison of ultracentrifugation, density gradient separation, and immunoaffinity capture methods for isolating human colon cancer cell line LIM1863-derived exosomes. Methods 56:293–304. https://doi.org/10.1016/j.ymeth.2012.01.002
8. Van Deun J, Mestdagh P, Sormunen R, Cocquyt V, Vermaelen K, Vandesompele J, Bracke M, De Wever O, Hendrix A (2014) The impact of disparate isolation methods for extracellular vesicles on downstream RNA profiling. J Extracell Vesicles 3:24858. https://doi.org/10.3402/jev.v3.24858
9. Kalra H, Adda CG, Liem M, Ang CS, Mechler A, Simpson RJ, Hulett MD, Mathivanan S (2013) Comparative proteomics evaluation of plasma exosome isolation techniques and assessment of the stability of exosomes in normal human blood plasma. Proteomics 13:3354–3364. https://doi.org/10.1002/pmic.201300282

10. Lane RE, Korbie D, Anderson W, Vaidyanathan R, Trau M (2015) Analysis of exosome purification methods using a model liposome system and tunable-resistive pulse sensing. Sci Rep 5:7639. https://doi.org/10.1038/srep07639
11. Alvarez ML, Khosroheidari M, Ravi RK, DiStefano JK (2012) Comparison of protein, microRNA, and mRNA yields using different methods of urinary exosome isolation for the discovery of kidney disease biomarkers. Kidney Int 82:1024–1032. https://doi.org/10.1038/ki.2012.256
12. Finoulst I, Pinkse M, Van Dongen W, Verhaert P (2011) Sample preparation techniques for the untargeted LC-MS-based discovery of peptides in complex biological matrices. J Biomed Biotechnol 2011. https://doi.org/10.1155/2011/245291
13. Granger J, Siddiqui J, Copeland S, Remick D (2005) Albumin depletion of human plasma also removes low abundance proteins including the cytokines. Proteomics 5:4713–4718. https://doi.org/10.1002/pmic.200401331
14. Yocum AK, Yu K, Oe T, Blair IA (2005) Effect of immunoaffinity depletion of human serum during proteomic investigations. J Proteome Res 4:1722–1731. https://doi.org/10.1021/pr0501721
15. Bellei E, Bergamini S, Monari E, Fantoni LI, Cuoghi A, Ozben T, Tomasi A (2011) High-abundance proteins depletion for serum proteomic analysis: concomitant removal of non-targeted proteins. J Amino Acids 40:145–156. https://doi.org/10.1007/s00726-010-0628-x
16. Capriotti AL, Caruso G, Cavaliere C, Piovesana S, Samperi R, Laganà A (2012) Comparison of three different enrichment strategies for serum low molecular weight protein identification using shotgun proteomics approach. Anal Chim Acta 740:58–65. https://doi.org/10.1016/j.aca.2012.06.033
17. Millioni R, Tolin S, Puricelli L, Sbrignadello S, Fadini GP, Tessari P, Arrigoni G (2011) High abundance proteins depletion vs low abundance proteins enrichment: comparison of methods to reduce the plasma proteome complexity. PLoS One 6(5):e19603. https://doi.org/10.1371/journal.pone.0019603
18. Zhao Y, Chang C, Qin P, Cao Q, Tian F, Jiang J, Li X, Yu W, Zhu Y, He F, Ying W, Qian X (2016) Mining the human plasma proteome with three-dimensional strategies by high-resolution Quadrupole Orbitrap Mass Spectrometry. Anal Chim Acta 904:65–75. https://doi.org/10.1016/j.aca.2015.11.001
19. Hakimi A, Auluck J, Jones GDD, Ng LL, Jones DJL (2014) Assessment of reproducibility in depletion and enrichment workflows for plasma proteomics using label-free quantitative data-independent LC-MS. Proteomics 14:4–13. https://doi.org/10.1002/pmic.201200563
20. Seong Y, Yoo YS, Akter H, Kang MJ (2017) Sample preparation for detection of low abundance proteins in human plasma using ultra-high performance liquid chromatography coupled with highly accurate mass spectrometry. J Chromatogr B 1060:272–280. https://doi.org/10.1016/j.jchromb.2017.06.023
21. Fernández C, Santos HM, Ruíz-Romero C, Blanco FJ, Capelo-Martínez JL (2011) A comparison of depletion versus equalization for reducing high-abundance proteins in human serum. Electrophoresis 32:2966–2974. https://doi.org/10.1002/elps.201100183
22. De Bock M, De Seny D, Meuwis MA, Servais AC, Minh TQ, Closset J, Chapelle JP, Louis E, Malaise M, Merville MP, Fillet M (2010) Comparison of three methods for fractionation and enrichment of low molecular weight proteins for SELDI-TOF-MS differential analysis. Talanta 82:245–254. https://doi.org/10.1016/j.talanta.2010.04.029
23. Liu G, Zhao Y, Angeles A, Hamuro LL, Arnold ME, Shen JX (2014) A novel and cost effective method of removing excess albumin from plasma/serum samples and its impacts on LC-MS/MS bioanalysis of therapeutic proteins. Anal Chem 86:8336–8343. https://doi.org/10.1021/ac501837t
24. Shi T, Fillmore TL, Gao Y, Zhao R, He J, Schepmoes AA, Nicora CD, Wu C, Chambers JL, Moore RJ, Kagan J, Srivastava S, Liu AY, Rodland KD, Liu T, Camp DG, Smith RD, Qian WJ (2013) Long-gradient separations coupled with selected reaction monitoring for highly sensitive, large scale targeted protein quantification in a single analysis. Anal Chem 85:9196–9203. https://doi.org/10.1021/ac402105s

25. Domon B, Aebersold R (2010) Options and considerations when selecting a quantitative proteomics strategy. Nat Biotechnol 28. https://doi.org/10.1038/nbt.1661
26. Picotti P, Aebersold R (2012) Selected reaction monitoring-based proteomics: workflows, potential, pitfalls and future directions. Nat Methods 9:555–566. https://doi.org/10.1074/mcp.M111.008235
27. Zhang Y, Fonslow BR, Shan B, Baek MC, Yates JR III (2013) Protein analysis by shotgun/ bottom-up proteomics. Chem Rev 113:2343–2394. https://doi.org/10.1021/cr3003533
28. Gillette MA, Carr SA (2013) Quantitative analysis of peptides and proteins in biomedicine by targeted mass spectrometry. Nat Methods 10:28–34. https://doi.org/10.1038/nmeth.2309
29. Gillet LC, Navarro P, Tate S, Röst H, Selevsek N, Reiter L, Bonner R, Aebersold R (2012) Targeted data extraction of the MS/MS spectra generated by data-independent acquisition: a new concept for consistent and accurate proteome analysis. Mol Cell Proteomics 11:1–17. https://doi.org/10.1074/mcp.O111.016717
30. Sajic T, Liu Y, Aebersold R (2015) Using data-independent, high-resolution mass spectrometry in protein biomarker research: perspectives and clinical applications. Proteomics Clin Appl 9:307–321. https://doi.org/10.1002/prca.201400117
31. Anjo SI, Santa C, Manadas B (2017) SWATH-MS as a tool for biomarker discovery: from basic research to clinical applications. Proteomics 17. https://doi.org/10.1002/pmic.201600278
32. Liu Y, Buil A, Collins BC, Gillet LC, Blum LC, Cheng LY, Vitek O, Mouritsen J, Iachance G, Spector TD, Dermitzakis ET, Aebersold R (2015) Quantitative variability of 342 plasma proteins in a human twin population. Mol Syst Biol 11:786. https://doi.org/10.15252/msb.20145728
33. Liu Y, Huettenhain R, Surinova S, Gillet LC, Mouritsen J, Brunner R, Navarro P, Aebersold R (2013) Quantitative measurements of N-linked glycoproteins in human plasma by SWATH-MS. Proteomics 13:1247–1256. https://doi.org/10.1002/pmic.201200417
34. Sanda M, Goldman R (2016) Data independent analysis of IgG glycoforms in samples of unfractionated human plasma. Anal Chem 88:10118–10125. https://doi.org/10.1021/acs.analchem.6b02554
35. Polaskova V, Kapur A, Khan A, Molloy MP, Baker MS (2010) High-abundance protein depletion: comparison of methods for human plasma biomarker discovery. Electrophoresis 31:471–482. https://doi.org/10.1002/elps.200900286
36. Drake RR, Schwegler EE, Malik G, Diaz J, Block T, Mehta A, Semmes OJ (2006) Lectin capture strategies combined with mass spectrometry for the discovery of serum glycoprotein biomarkers. Mol Cell Proteomics 5:1957–1967. https://doi.org/10.1074/mcp.M600176-MCP200
37. Gundry RL, White MY, Nogee J, Tchernyshyov I, Van Eyk JE (2009) Assessment of albumin removal from an immunoaffinity spin column: critical implications for proteomic examination of the albuminome and albumin-depleted samples. Proteomics 9:2021–2028. https://doi.org/10.1002/pmic.200800686
38. Govorukhina NI, Keizer-Gunnink A, Van der Zee AGJ, De Jong S, De Bruijn HWA, Bischoff R (2003) Sample preparation of human serum for the analysis of tumor markers: comparison of different approaches for albumin and g-globulin depletion. J Chromatogr A 1009:171–178. https://doi.org/10.1016/S0021-9673(03)00921-X
39. Zolotarjova N, Martosella J, Nicol G, Bailey J, Boyes BE, Barrett WC (2005) Differences among techniques for high-abundant protein depletion. Proteomics 5:3304–3313. https://doi.org/10.1002/pmic.200402021
40. Théry C, Amigorena S, Raposo G, Clayton A (2006) Isolation and characterization of exosomes from cell culture supernatants and biological fluids. Curr Protoc Cell Biol:3–22. https://doi.org/10.1002/0471143030.cb0322s30
41. Ko J, Carpenter E, Issadore D (2016) Detection and isolation of circulating exosomes and microvesicles for cancer monitoring and diagnostics using micro-/nano-based devices. Analyst 141:450–460. https://doi.org/10.1039/C5AN01610J
42. Taylor DD, Gercel-Taylor C (2008) MicroRNA signatures of tumor-derived exosomes as diagnostic biomarkers of ovarian cancer. Gynecol Oncol 110:13–21. https://doi.org/10.1016/j.ygyno.2008.04.033

43. Skotland T, Sandvig K, Llorente A (2017) Lipids in exosomes: current knowledge and the way forward. Prog Lipid Res 66:30–41. https://doi.org/10.1016/j.plipres.2017.03.001
44. Record M, Carayon K, Poirot M, Silvente-Poirot S (2014) Exosomes as new vesicular lipid transporters involved in cell – cell communication and various pathophysiologies. Biochim Biophys Acta 1841:108–120. https://doi.org/10.1016/j.bbalip.2013.10.004
45. Muller L, Hong CS, Stolz DB, Watkins SC, Whiteside TL (2014) Isolation of biologically-active exosomes from human plasma. J Immunol Methods 411:55–65. https://doi.org/10.1016/j.jim.2014.06.007
46. Turay D, Khan S, Diaz Osterman CJ, Curtis MP, Khaira B, Neidigh JW, Mirshahidi S, Casiano CA, Wall NR (2016) Proteomic profiling of serum-derived exosomes from ethnically diverse prostate cancer patients. Cancer Investig 34:1–11. https://doi.org/10.3109/07357907.2015.108 1921
47. Ahmed FE (2009) Sample preparation and fractionation for proteome analysis and cancer biomarker discovery by mass spectrometry. J Sep Sci 32:771–798. https://doi.org/10.1002/jssc.200800622
48. Zhang L, Wang W, Bai J, Xu YF, Li LQ, Hua L, Deng L, Jia H (2016) Proteomic analysis associated with coronary artery dilatation caused by Kawasaki disease using serum exosomes. Rev Port Cardiol 35:265–273. https://doi.org/10.1016/j.repc.2015.11.016
49. Kowal J, Arras G, Colombo M, Jouve M, Morath PJ, Primdal-Bengtson B, Dingli F, Leow D, Tkach M, Théry C (2016) Proteomic comparison defines novel markers to characterize heterogeneous populations of extracellular vesicle subtypes. Proc Natl Acad Sci 113:968–977. https://doi.org/10.1073/pnas.1521230113
50. Smith MPW, Wood SL, Zougman A, Ho JT, Peng J, Jackson D, Cairns DA, Lewington AJP, Selby PJ, Banks RE (2011) A systematic analysis of the effects of increasing degrees of serum immunodepletion in terms of depth of coverage and other key aspects in top-down and bottom-up proteomic analyses. Proteomics 11:2222–2235. https://doi.org/10.1002/pmic.201100005
51. Dayon L, Kussmann M (2013) Proteomics of human plasma: a critical comparison of analytical workflows in terms of effort, throughput and outcome. EuPA Open Proteom 1:8–16. https://doi.org/10.1016/j.euprot.2013.08.001
52. Zhao X, Wu Y, Duan J, Ma Y, Shen Z, Wei L, Cui X, Zhang J, Xie Y, Liu J (2014) Quantitative proteomic analysis of exosome protein content changes induced by hepatitis B virus in Huh-7 cells using SILAC labeling and LC−MS/MS. J Proteome Res 13:5391–5402. https://doi.org/10.1021/pr5008703
53. An M, Lohse I, Tan Z, Zhu J, Wu J, Kurapati H, Morgan MA, Lawrence TS, Cuneo KC, Lubman DM (2017) Quantitative proteomic analysis of serum exosomes from patients with locally advanced pancreatic cancer undergoing chemoradiotherapy. J Proteome Res 16:1763–1772. https://doi.org/10.1021/acs.jproteome.7b00024
54. Baranyai T, Herczeg K, Onódi Z, Voszka I, Módos K, Marton N, Nagy G, Mäger I, Wood MJ, El Andaloussi S, Pálinkás Z, Kumar V, Nagy P, Kittel A, Buzas EI, Ferdinandy P, Giricz Z (2015) Isolation of exosomes from blood plasma: qualitative and quantitative comparison of ultracentrifugation and size exclusion chromatography methods. PLoS One 10:e0145686. https://doi.org/10.1371/journal.pone.0145686

Chapter 4
Sample Treatment for Tissue Proteomics in Cancer, Toxicology, and Forensics

L. M. Cole, M. R. Clench, and S. Francese

4.1 Tissue Preparation and Treatment in Oncological Proteomics

The preparation and pretreatment of tissues for the study of proteomics in cancer are both unique and challenging considering the information sought by the clinician or research scientist. This is true regardless of sample origin, whether this be a patient clinical biopsy, from experimental animal models or novel synthetic tissues used in the field of oncology. Preparation is a key factor in order to extract maximum protein/peptide yield to ultimately generate that 'characteristic molecular snapshot', a true indication of proteomic involvement in the pathway to disease.

This account aims to discuss various perspectives of sample preparation and pretreatment of human tumor tissue samples essential to study proteomics using advanced analytical techniques. The methodologies and considerations herein are applicable to the techniques of matrix-assisted laser desorption/ionization mass spectrometry profiling and imaging (MALDI-MSP and MALDI-MSI, respectively) and liquid extraction surface analysis (LESA).

Fresh Frozen Tissue There are many challenges and factors to consider when selecting fresh frozen (FF) tissue as the storage method of choice prior to sample analysis. Experimental design that involves FF human tissue must not only account for ethical approval but sample availability, sample handling, and also the duration between sample harvest and snap freezing, a factor where time really is of the essence.

L. M. Cole · M. R. Clench · S. Francese (✉)
Biomolecular Science Research Centre, Centre for Mass Spectrometry Imaging, Sheffield Hallam University, Sheffield, UK
e-mail: s.francese@shu.ac.uk

© Springer Nature Switzerland AG 2019   77
J.-L. Capelo-Martínez (ed.), *Emerging Sample Treatments in Proteomics*, Advances in Experimental Medicine and Biology 1073, https://doi.org/10.1007/978-3-030-12298-0_4

In order to generate reproducible data from FF tissue that is believed a true, accurate proteomic snapshot, rigorous measures will need to be adhered to minimize potential proteolytic degradation. Ex vivo tissue will still continue to undergo changes in molecular integrity after removal from the site of biopsy. Such post-sampling will result in ischemic effects which could induce apoptosis and in situ coagulation, all conditions which encourage protein degradation [1].

Good technical skills, laboratory procedures, and storage facilities will collectively aid in the preservation of molecular integrity of FF samples for future proteomic analysis. Unlike paraffin-embedded tissue that has various stages to tissue preparation (wax removal and antigen retrieval), if good laboratory practices are carried out and the FF tissue is treated carefully, there should be minimal loss of target species.

The main advantage of analyzing molecular targets from human tissue in general could be the potential of translational, even personalized, medicine that could help advance disease diagnostics.

Tryptic Peptide Analysis One of the main potential pitfalls of in situ tissue tryptic digestion is protein delocalization, a factor which, when occurs, prevents the analysis of spatial protein distribution. To ensure good tryptic digest MALDI-MSI, a homogeneous coating is essential during trypsin application. Other issues to consider would be the tissue type, the section size, and the selection of a suitable flow rate, when using a pneumatic sprayer for trypsin deposition. Too much trypsin sprayed per cycle could lead to "over-wetting" of the tissue and subsequent protein delocalization. An example of a successful *on-tissue* tryptic digest can be seen in Fig. 4.1.

A parameter not always considered during enzyme application to a sample like tumor tissue is not only the 'x and y' uniformity but the depth and degree of enzyme seepage within the 'z' plane of the sample. This other 'z' dimension could also reflect how efficient an *on-tissue* is or on the other hand be indicative of "noisy" spectra due to degradation products of an overdigested sample.

A summary of issues and influential factors that potentially effect in situ tissue tryptic digestion and MALDI-MSI is presented in Fig. 4.2.

The general molecular heterogeneity of tissue samples means that it is vital to maintain the spatial integrity of proteins, especially post-enzymatic digestion. There are numerous studies that have been performed using mass spectrometric techniques on fresh frozen tissues, each study contributing a novel addition to the era of tissue proteomics using mass spectrometry [2–5].

As with all published research in a similar scientific field, each study critically evaluates another body of research against their own data output in order to optimize and progress methodologies. A two-center study performed in 2014 held in Leiden, Netherlands, and in Munich, Germany, did one such comparison. Both groups embarked on a comparative set of experiments and produced the output entitled "Multicenter Matrix-Assisted Laser Desorption/Ionization Mass Spectrometry Imaging (MALDI-MSI) Identifies Proteomic Differences in Breast-Cancer-Associated Stroma" [6]. The groups aimed to undertake a '*center-to-center*'

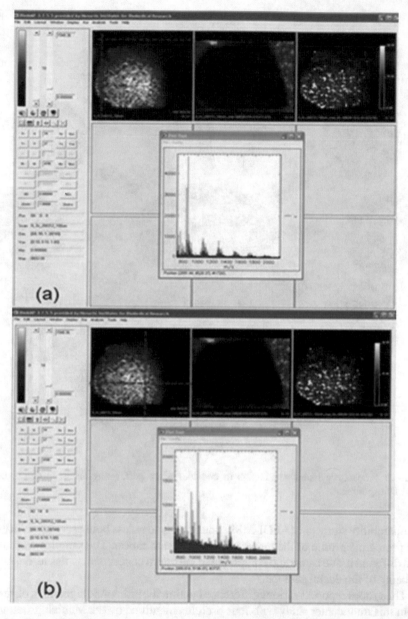

Fig. 4.1 Screenshots of a successful *on-tissue* tryptic digest using MALDI-MSI, visualized using BioMap imaging software. Screenshot (**a**) displays the corresponding spectrum in relation to the position of the red cross hairs which lie in a region that should be predominantly MALDI matrix. Similarly screenshot (**b**) indicates the peptide mass fingerprint typical of an *on-tissue* tryptic digest, with the red cross hairs in the digested tumor tissue region

Fig. 4.2 Preparatory issues and factors to consider when performing on-tissue digestion for MALDI experiments

comparability study of MALDI-MSI experiments to assess both user variability and the proteomic profile of the tumor stroma in breast cancer. The samples were from two different tissue banks and comprised of FF breast cancer patient tissue, described as being of the ductal subtype.

The groups reported various differences within the proteomic signatures observed from this multicenter study with four proteins identified by the Munich group to be significantly associated with the tumor stromal region and three of these signals also evident in the Leiden data set. The complex MALDI-MSI data produced was validated using immunohistochemical staining and a range of statistical tests including hierarchical cluster analysis.

It is reported within the article that dissimilarities were expected due to the differing sample treatment methods, some of which are detailed within Table 4.1.

Table 4.1 Methodologies employed in the two-center study between Leiden and Munich

	Leiden	Munich
Tissue washing steps	60% methanol (1 min), 100% methanol (1 min)	70% ethanol (1 min), 100% ethanol (1 min)
Matrix (application system)	Sinapinic acid (ImagePrep device)	Sinapinic acid (ImagePrep device)
Mass spectrometer (mode)	UltrafleXtreme (positive linear)	Ultraflex III (positive linear)
m/z range	2000–28,000	2520–25,100
Lateral resolution	150 μm	70 μm
Hematoxylin and eosin co-registration	Same slide	Same slide
Maximum peak shift (ppm)	1000	1000
% match to calibrant peaks	20	20

Adapted from [6]

Research scientists often use a range of multidisciplinary workflows to aid the validation of proteomic biomarkers. Frozen samples do not exclusively describe a mounted tissue section and increasingly mass spectrometry is being used for needle biopsy analysis. MALDI-MSI has been used within workflows to assess proteins and match morphological with molecular data in cytological smears [7]. Following a top-down study by Pagni et al. (2016) [7], zones of interest from thyroidectomy specimens were preselected by a pathologist (using a May-Grünwald-Giemsa-stained image) in order for the mass spectrometrist to perform a virtual microdissection of that region of interest post-MALDI-MSI. The spectra within the area chosen were then exported for data analysis. The proteins observed could potentially be correlated to malignant or benign regions of the specimen. The tissue preparations detailed within the publication were minimal due to the top-down nature of the experiments. Prior to matrix deposition, the sample slides were thawed, desiccated (30 min), then fixed, and washed with ethanol followed by another desiccation step (10 min). Sinapinic acid (10 mg/mL in 60%/0.2% v/v acetonitrile/TFA) was applied using the ImagePrep (Bruker Daltonik, GmbH, Bremen, Germany).

Ensuring that experimental methods and techniques deliver maximum data yield is vital and the opportunity to reuse and diversify samples would allow flexibility in a workflow and reap financial benefits. Yet, does this ideal exist in tissue proteomics?

A recent study performed by Randall et al. (2016) presented a combination of MALDI and LESA for the investigation of lipids, drugs, and proteins all from a single tissue section [8]. This collaborative study used animal brain tissue but, the methodologies could potentially be translational to human tissue samples. The analysis indicated how sampling with LESA altered the corresponding MALDI spectra and integrity of the tissue sample. To help define such changes, the group employed multivariate statistical models to assess each sampling point post-LESA and subsequent molecular yield. To perform liquid extraction surface analysis mass spectrometry, pipette "z coordinate" was set to a height 0 mm; by doing this the sampling tip came into contact with the tissue surface rather than a noncontact microjunction method, which is usually set to ~0.2 mm (using the TriVersa NanoMate chip-based electrospray device (Advion, Ithaca, NY)). The solvent for liquid extraction used for LESA MS imaging was methanol, water, and formic acid (69.3:29.7:1). MALDI-MSI post-LESA sampling was then performed after matrix deposition.

Although LESA imaging has its own limitations in terms of spatial resolution and the ever-increasing need for smaller pixel size, the perception that a single tissue section can be used for multiple analyses such as drug-lipid-protein sampling is a very appealing prospect indeed.

Formalin-Fixed Paraffin-Embedded Tissue Preparing FFPE tissues for mass spectrometry profiling and/or imaging presents many challenges, some of which include detachment of tissue from the microscope slide, interference from remaining paraffin wax, delocalization, and loss of target species of interest. These factors affect reproducibility and increase the requirement for biological replicates in addition to technical repeats. However, tissues from patient biopsies are often embedded in paraffin after formalin fixation either for immediate histological analysis or retained for tissue bank archiving. The formalin-fixed paraffin-embedded (FFPE) tissue protocol has been extensively used for tissue preservation based on its methylene bridge cross-linking properties [9–11]. This method allows room temperature storage of patient tissue enhancing sample longevity and storage flexibility.

There are studies that have been performed on sample archival time, the impact of sample 'block age,' and potential impact of sample variability. Craven et al. (2013) performed one such study on patient renal tissue for the application of LC-MS/MS using a label-free, bottom-up approach [12]. These data generated proteins using historical FPPE archived tissues dating from 2001 to 2011. Interestingly, block age was not found to be the limiting factor in terms of peptide yield and protein hits but established the importance of tumor grading to inform and reflect disease heterogeneity. This study considered not only experimental design but many influential factors that could impact on downstream results. Sample type and histopathology prior to mass spectrometry experiments were defined with factors that included age, gender, hemorrhagic appearance, necrosis, and features indicating inflammation. With around 2000 proteins identified from each sample, the results provided reassurance that archival tissues can be analyzed to assess proteomics of diseases like renal cell carcinoma and ultimately yield clinically relevant data.

In addition to pre-experimental histological selection, the tissue preparations employed in the above Craven et al. (2013) study can be transferrable to other mass spectrometry techniques, i.e., MALDI and LESA. The methods describe how 10 μm tissue sections were cut and dewaxed in the standard way (xylene and hydration varying ethanol concentrations). After these stages, a 5 cm^2 area from the sample was macrodissected using a scalpel, with stained hematoxylin and eosin (H&E) serial tissue sections providing areas of interest included. This histological staining as described here could be employed to help focus and reveal regions of interest prior to performing LESA experiments. The next steps in this proteomics experiment would involve some form of enzymatic digestion in a bottom-up approach, i.e., to observe tryptic peptides. For LESA and MALDI-MSP, applications direct from the tissue surface and then the methods described earlier in the 'Fresh Frozen Tissue' section of this account would be applicable in terms of the enzyme application.

To be applicable to MALDI-MSP, the material from the macrodissection, once in lysis buffer solution (62.5 mM Tris-HCl, pH 6.8, 4% w/v SDS, 100 mM DTT, 10% v/v glycerol), could be treated using the following described preparations even though originally designed for LC-ESI-MS/MS (ESI – electrospray ionisation); the sample protocol that Craven et al. (2013) employed was extensive, but, in doing this, their efforts resulted in being able to claim that block age did not correlate to a decreased protein yield.

Post-sample macrodissection, the group reported that lysates were incubated at 105 °C for 45 min and then on ice (5 min) before being pushed through a syringe in order to shear DNA. Subsequently, the solution was microfuged for 10 min at 4 °C and then, to minimize sample degradation, stored at −80 °C until required. To perform tryptic digestion, lysates were heated (95 °C, 2 min) after which the group reported that a urea (UA) buffer solution (50 mM ammonium bicarbonate containing 8 M urea) containing ~250 µg of protein was then digested by filter-aided sample preparation as performed by previous studies [13, 14]. Filter-aided sample preparation (FASP) describes a method where anion exchange pipette tip columns are used. The sequential steps that are detailed in Fig. 4.3 describe the additions in solution preparation during FASP; the flow-through was discarded after any centrifugation steps were introduced.

Immunohistochemical staining of FFPE patient samples is routinely performed within the clinical setting and within research to observe protein targets in both control and diseased tissue. To unlock and enable access to such biomarker targets within an excised tissue biopsy, the process of antigen retrieval (AR) is deemed a crucial stage in sample treatment. This reversal of the protein cross-linking process induced by FFPE can be considered a key factor in maximizing peptide yield,

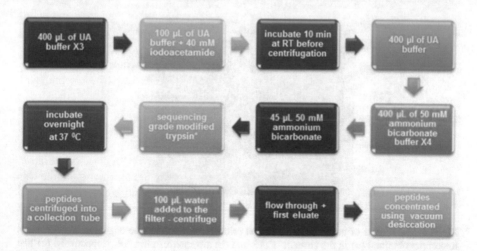

Fig. 4.3 This workflow details the sequential steps used in a FASP experiment to yield tryptic peptides for proteomic analysis. (∗20 µg of lyophilized trypsin was resuspended in 80 µL 50 mM ammonium bicarbonate buffer-to-protein ratio of 1:50)

thus improving the volume and quality of signals from subsequent peptide mass fingerprints.

Djidja et al. (2017) published a study within this developmental field of tissue AR treatment using ex vivo human breast tumor tissue preserved by the FFPE method [15]. The resulting peptide yield in the in situ digestion spectra was quite profound after performing AR compared with that from in situ digestion in the absence of AR (see Fig. 4.4 (a) and (b), respectively). To aid explanation of the AR data further, a graphical representation (Fig. 4.4 (c)) indicates the average peak intensities of peptide ion signals from peptides belonging to actin (m/z 1198.7) and retinoic acid early transcript 1G protein (m/z 1095.6), with and without the AR step in the tissue treatment. It is also worth mentioning that the subsequent protein identifications (see Table 4.2) were remarkable, considering that these data were

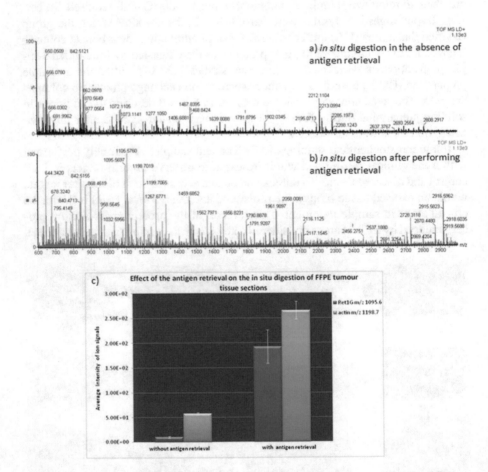

Fig. 4.4 The effects of using antigen retrieval prior to *on-tissue* tryptic digestion. Peptide mass fingerprint obtained (**a**) in the absence of AR and (**b**) after performing AR, respectively. (**c**) The average peak intensities of peptide ion signals of actin (m/z 1198.7) and retinoic acid early transcript 1G protein (m/z 1095.6) are shown in the absence and use of AR prior to in situ digestion (n = 3). (Reprinted from [15], Copyright (2017) with permission from Elsevier)

Table 4.2 Tryptic peptides that were identified after in situ digestion of FFPE breast tumor tissue sections using direct MALDI-MS/MS

Protein name/accession number	Mass (Da)	Observed m/z	Mass error (ppm)	Sequence	MOWSE score	Location and function
Actin, cytoplasmic 1/ P60709	41,710	976.43	−13.5	AGFAGDDAPR	21	Cytoplasm; cell mobility
		1198.69	−9.60	AVFPSIVGRPR	12	
Collagen alpha-1 (I)/ P02452	138,827	836.43	−3.75	GPAGPQGPR	55	Extracellular matrix, secreted.
		852.43	−0.76	GPPGPQGAR, oxidation (P)	31	Fibril organization, blood vessel
		1105.57	−8.06	GVQGPPGPAGPR, oxidation (P)	24	development
		1546.79	−0.49	GETGPAGPAGPVGPVGAR	75	
		1585.76	−4.98	GANGAPGIAGAPGFPGAR, 3 oxidation (P)	34	
		1775.87	2.8	GARGEPGPTGLPGPPGER, acetyl (N-term); 2 oxidation (P)	11	
Collagen alpha-2 (I)/ P08123	129,209	868.45	−18.3	VGAPGPAGAR, oxidation (P)	14	
		960.45	−11.7	AGVMGPPGSR, oxidation (M); Oxidation (P)	16	
		1562.79	−1.9	GETGPSGPVGPAGAVGPR	26	
		1580.73	−19.9	GPPGESGAAGPTGPIGSR, oxidation (P)	79	
		1751.80	−16.2	GPPGAVGSPGVNGAPGEAGR, 3 oxidation (P)	50	
		1775.87	−3.52	RGPNGEAGSAGPPGPPGLR, 2 oxidation (P)	39	
		1833.89	3.97	RGPNGEAGSAGPPGPPGLR, Acetyl (N-term); 3 oxidation (P)	28	

(continued)

Table 4.2 (continued)

Protein name/accession number	Mass (Da)	Observed m/z	Mass error (ppm)	Sequence	MOWSE score	Location and function
Collagen alpha-1 (III)/ P02461	138,479	836.43	−3.75	GAPGPQGPR	43	
		852.43	−0.76	GPPGPQGAR, oxidation (P)	25	
		1094.58	−9.87	GPPGLAGAPGLR, 2 oxidation (P)	37	
		1111.59	−4.6	GRPGLPGAAGAR, 2 oxidation (P)	41	
		1154.56	−1.11	GLAGPPGMPGPR, oxidation (M); 2 oxidation (P)	33	
		1303.61	4.85	GSPGGPGAAGFPGAR, 3 oxidation (P)	48	
		1320.66	−7.12	GPPGPAGANGAPGLR, 2 oxidation (P)	18	
		1508.72	9.67	GESGPAGPAGAPGPAGSR, oxidation (P)	20	
		1702.79	2.81	GEMGPAGIPGAPGLMGAR, 2 oxidation (M); 2 oxidation (P)	13	
		1833.89	−9.76	GPPGPQGLPGLAGTAGEPGR, 3 oxidation (P)	12	
		2104.07	−3.70	GSPGAQGPPGAPGPLGIAGITGAR, 3 oxidation (P)	25	
Collagen alpha-1 (X)/ Q03692	66,117	1466.67	−9.59	GLNGPTGPPGPPGPR, 6 oxidation (P)	21	
Heat-shock protein beta-1/P04792	22,768	987.61	−1.22	RVPFSLLR	26	Cytoplasm, nucleus; involved in stress resistance
		1163.61	7.54	LFDQAFGLPR	31	

Histone H2AZ/P0C0S5	13,545	944.53	−5.55	AGLQFPVGR	41	Nucleus; gene regulation of
Histone-lysine	563,831	1095.55	0.21	GGAHGGRGRGR, acetyl (N-term); oxidation (HW)	14	nucleus, acts as a coactivator for estrogen receptor, activates transcription
N-methyltransferase/ O14686		1286.64	−6.24	NLTMSPLHKR, acetyl (N-term); oxidation (HW); oxidation (M); oxidation (P)	23	
Keratin, type II cytoskeletal 7/P08729	51,386	1406.68	−18.8	SIHFSSPVFTSR, acetyl (N-term)	40	Cytoplasm; structural molecule
Obscurin-like protein 1/ O75147	152,786	1111.59	5.44	NGAVVTPGPQR, oxidation (P)	33	Muscle development

Reprinted from [15], Copyright (2017) with permission from Elsevier acquired by means of MALDI-MS/MS directly from FFPE breast tumor tissue sections, hence omitting a pre-separation step. Furthermore, de novo sequencing revealed proline hydroxylation, and, as a result, the authors report that this may be due to the occurrence of chemical derivatization introduced during AR [16].

After the conventional paraffin removal and hydration steps using xylene and graded ethanol, the AR method included the immersion of tissue in a hydrogen peroxide solution (3% H_2O_2 in methanol, 12 min at room temperature) [15]. The latter treatment was the speculative proline hydroxylation stage mentioned above, knowing of the potentiation of proline to quench hydroxyl radicals therefore initiating diverse compounds such as hydroxyproline to arise [17, 18]. Pre-enzymatic digestion, the samples were placed in tri-sodium citrate buffer (at 0.01 M, at pH 6.3) and heated in a microwave oven (13 min until simmering). After which the tissue sections were left to equilibrate to room temperature prior to trypsin and matrix deposition.

In the quest to maintain a true insight into the proteomic signature of a diseased tumor section, or likewise the ability to screen the chemical biology from patient excised tissue, it is imperative to encourage the momentum of research into molecular target availability. Although MALDI matrix sublimation is now a widespread technique, one such study by Lin et al. (2018) aimed to push the boundaries of molecular extraction one step further [19]. As the title of the application note describes, their method of matrix deposition by sublimation claims to enhance MALDI-MSI by introducing a simple sonication step to improve proteomic signals. The authors report that an additional step of hydration postmatrix deposition considerably improves ionization efficiency of proteins. The article is based on the fact that sublimation aims to minimize matrix 'hot spots' and promote a homogeneous coating, desirable for imaging mass spectrometry experiments [20, 21]. In addition to this, the publication also recognized that sublimation is notoriously poor for efficient protein extraction. After sublimation and hydration stages adhering to protocols similar to those used by previous studies, sonication of the sample sealed in a petri dish chamber occurred at 37 kHz for either 2 or 5 min. Differences in the temperature between sonication parameters were recorded; after 2 min the change in temperature increased from 24 to 26 °C and to 30 °C after 5 min of sonication. Interestingly, they conclude an improvement in signal to noise particularly in the higher mass m/z range (>m/z 5000), with 2 min being the optimal sonication duration. The results showed differences in signal intensity from protein to protein with effects of sonication providing no significant advantage up to 5 min.

The proposed explanation for the improved signal post-sonication was described to be in alignment with previous studies including findings by Je et al. (2006) [22]. The rationale suggested was that the hydration step alone post-sublimation could extract small and hydrophilic species allowing cocrystallization. Conversely, sonication permitted mobilization of larger proteins, thus allowing fair competition for matrix crystal free surface. The publication also describes an analogy that envisages a microjet mixing of the liquid-solid interface due to the ultrasonic wave pressures.

Tissue microarrays (TMA) can be regarded as being one step closer to achieving '*higher-throughput*' mass spectrometry imaging. Djidja et al. (2010) previously reported this proof of concept nearly a decade ago where the group employed MALDI-MSI to observe peptides in a TMA tumor classification study using human pancreatic tissue cores [23]. The TMA imaged in this study comprised of 60 adenocarcinoma tissue cores (30 patients in duplicate). To ensure proteins were not delocalized during enzymatic application, as mentioned previously in the "Fresh Frozen Tissue" section, a series of trypsin layers were pneumatically sprayed (SunCollect™ automatic sprayer, SunChrom, Friedrichsdorf, Germany); the first layer of trypsin was a flow rate 1 μL/min and the second at 2 μL/min and the final three layers were sprayed at 4 μL/min. The group also used a similar order of layering to deposit the MALDI matrix postdigestion.

Recently, a TMA study but on a larger scale was performed by Kriegsmann et al. (2017), looking at the differentiation and proteomic molecular signatures of adenocarcinomas, colon versus lung cancer [24]. In this study, 383 FFPE tissue specimens were analyzed from individual patients with primary adenocarcinoma. The cohorts included adenocarcinoma of the colon ($n = 217$) and primary adenocarcinoma of the lung ($n = 166$). Although on a larger scale, the tissue treatments before mass spectrometric experimental work are comparable and only differ by the trypsin and matrix deposition instrumentation, that being the Image Prep (Bruker Daltonik, GmbH, Bremen, Germany) in this study.

4.2 Sample Preparation for Tissue Proteomics in Toxicology

The application of proteomics to studies in toxicology has been termed "toxicoproteomics." In the following section, the application of and methodologies applied in toxicoproteomics will be illustrated with examples from its use in two important areas: the study of drug-induced liver injury (DILI) and studies of effects of chemical and environmental insults on skin, i.e., the effects of irritants, sensitizers, and ionizing radiation.

4.2.1 Drug-Induced Liver Injury

Drug-induced liver injury (DILI) is defined as liver injury caused by drugs, leading to abnormalities in liver tests or liver dysfunction with the reasonable exclusion of other causes [25]. DILI is one of the leading causes of acute liver failure, accounting for 13% of cases of acute liver failure in the USA. DILI is thought to occur via a number of mechanisms: direct impairment by the drug itself of the structure and functional integrity of the liver; production of a drug biotransformation product (metabolite) that alters liver structure and function; production of a reactive drug metabolite that binds to hepatic proteins to produce new antigenic drug-protein adducts (which are targeted by immunological defenses); and initiation of a systemic hypersensitivity response (i.e., an allergic response) that damages the liver. DILI poses a major challenge to pharma for drug development and safety. Alanine aminotransferase, aspartate aminotransferase, alkaline phosphatase, and total bilirubin are currently the only approved DILI biomarkers in clinical practice. While fulfilling an important role in disease diagnostics, these biomarkers are not specific for hepatotoxicity, as their levels increase in practically all liver conditions [26]. Therefore, in order to find more specific biomarkers, proteomics-based studies of DILI are ongoing.

It has been postulated that DILI is influenced by underlying environmental and genetic factors as well as the chemical properties of the administered drug. This was the subject of a study conducted in 2015 by Ramm et al. [27]. The study provides an excellent illustration of the use of proteomics in the study of DILI. An initial LC-MS/MS proteomics screen followed by a targeted LC-MS/MS (MRM – multiple reaction monitoring) proteomics approach was used to identify liver and plasma proteins modulated by bacterial endotoxin (LPS), diclofenac (Dcl), or LPS/Dcl cotreatment. The overall toxicoproteomics strategies employed are shown in Fig. 4.5.

Data from the initial screening experiment carried out label-free in data-dependent acquisition (DDA) mode on a quadruple time-of-flight (QqTOF) instrument was analyzed using Progenesis QIP software (nonlinear dynamics) and the results combined with those from previous studies and the literature to select 47 proteins for MRM-based quantification. The quantification experiments were carried out on a QqQ (triple quadrupole) instrument with unique peptides for quantification selected using Skyline (https://skyline.ms) and three transitions per peptide used in the final experiment.

In order to prepare the liver samples for proteomic analysis, a protocol was adopted which is widely applicable and is illustrated in Table 4.3.

The experimental protocol detailed above can be taken as typical of those used for studies of this nature and gives a good starting point for any proposed work.

Acetaminophen (paracetamol) is the drug that has been most widely studied in DILI. Its effects have been studied in their own right and it has also been used as a model system for DILI studies. An interesting recent paper [28] reports the study of acetaminophen-induced hepatotoxicity carried out using $^{18}O/^{16}O$ labelling for quantitative proteomics. Here comparison of liver tissue isolated from low (100 mg/kg) and high (1250 mg/kg) dose rats was performed. While the methodology employed for the generation of protein digests for LC/MS/MS is comparable to that used by Ramm et al. [27], it is the use of $^{18}O/^{16}O$ labelling for quantitative comparison that is of interest. A detailed protocol for preparing samples for $^{18}O/^{16}O$ labelling has been published by Castillo et al. [29]. In summary protein extracts for comparison are enzymatically digested in parallel in $H_2^{16}O$ (i.e., standard deionized water) and $H_2^{18}O$ (^{18}O-labelled water). Since the digest is a hydrolysis reaction,

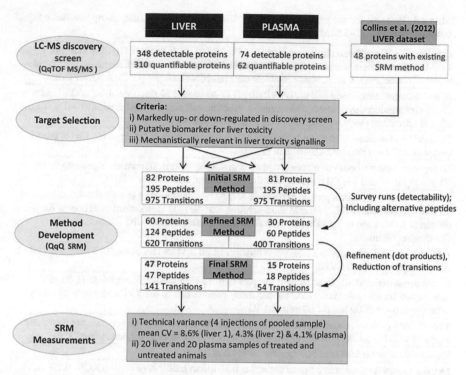

Fig. 4.5 Toxicoproteomics method development workflow for the study of DILI. The proteins selected for monitoring by MRM were derived from an initial label-free LC-MS discovery proteomics screen of five liver and five plasma samples (pooled), a set of liver proteins previously studied, and examination of the literature on potential candidate hepatotoxicity biomarkers. The initial MRM transition list was evaluated by injection of one liver and one plasma sample (pooled) to determine the detectability of peptides mapping to proteins of interest. Based on this data, a refined MRM method was built, including alternative peptides for the so far undetectable proteins. For the final MRM method, peptide signals of good quality (determined by comparison with library MS/MS spectra: dot product calculation) were kept and the number of transitions for each peptide reduced from 5 to 3. The optimized MRM liver (split in two) and plasma methods were then applied to measurements in entire sample set. (Reprinted from [27], Copyright (2015) with permission from Elsevier)

in the digest carried out in ^{18}O water, two atoms of ^{18}O are incorporated into the carboxyl terminus of each peptide generated in the digest. Comparative proteomic studies are performed by mixing the sample digest containing unlabelled peptides (generated in normal water) and with digest performed in the isotope-labelled water which contains ^{18}O peptides and analyzing the resultant peptide pairs by mass spectrometry.

Relative quantification is derived from ratios of the isotope pairs. Conventional MS/MS experiments can then be used to identify the proteins. Given the relative simplicity of this procedure it is impressive that approximately 1000 proteins were identified and quantified in each sample studied. The overall outcome of

Table 4.3 Protocol for the preparation of liver tissue for proteomic analysis by nanoscale liquid chromatography (nLC)-MS/MS

Homogenize 50 mg frozen liver tissue on ice with 120 µL lysis buffer I (10 mM Tris (pH 7.5), 1 mM ethylenediaminetetraacetic acid (EDTA), 0.2 M sucrose, 1.25 U Benzonase per 120 µL buffer (Calbiochem), 1 µL protease inhibitor cocktail (Calbiochem) per 20 mg liver tissue)
After homogenization, add 880 µL lysis buffer II (7 M urea, 2 M thiourea, 4% (w/v) CHAPS (3-[(3-cholamidopropyl)dimethylammonio]-1-propanesulfonate), 4 mM dithiothreitol (DTT), 20 mM spermine) (and re-pipette 30 times to aid suspension)
Incubate protein extracts at room temperature on a rotary shaker at 500 rpm for 1 h (to ensure complete cell lysis and solubilization of proteins)
To separate membranous components and other insoluble debris, ultracentrifuge samples for 30 min at 10 °C and 74,000 × g. Aliquote the supernatant and store at −80 °C
Denature liver and plasma protein extracts (200 mg) by resuspending to a final concentration of 50% trifluoroethanol (TFE), 10 mM dithiothreitol, and 50 mM ammonium bicarbonate for 60 min at room temp in a 5 kDa Mw spin filter device (Sartorius Stedim Biotech GmbH, Goettingen, Germany)
Add iodoacetamide (IAA) to a final concentration of 20 mM and incubate for a further 60 min at room temperature in the dark
Dilute denatured and alkylated protein samples tenfold with wash buffer (5% TFE, 50 mM ammonium bicarbonate buffer) and subsequently concentrate in the 5 kDa Mw spin filters by centrifugation at 3000 × g for 60 min at 4 °C
The retained protein containing samples should now be transferred to a separate container and digested with the addition of 0.2 µg/µL trypsin to achieve a protein/trypsin ratio of 1:50
Incubate in a thermomixer at 500 rpm for 18–24 h at 37 °C
Dry the digested samples in a SpeedVac and resuspend in buffer A (3% (v/v) ACN, 0.1% (v/v) formic acid) to 0.5 µg/µL prior to nLC-MS/MS

Adapted from Ramm et al. [27]

this study was the identification of heme oxygenase 1 as a potential plasma biomarker of DILI.

In twenty-first-century toxicology, there is a great deal of interest in moving away from conventional animal-based studies to in vitro experiments [30]. For the study of DILI acetaminophen-treated three-dimensional (3D) liver microtissues were employed in an interesting study reported by Bruderer et al. [31]. The liver microtissues used were a coculture of consisting of primary human hepatocytes and primary human nonparenchymal cells. These were obtained commercially from InSphero AG (Switzerland) as 3D InSight™ Human Liver Microtissues. For the preparation of the microtissues, the protocol employed is summarized in Table 4.4.

In this study the mass spectrometric methodology employed was a novel data-independent acquisition (DIA) method for hypermonitoring (HRM) (https://biognosys.com).

Due to regulatory, economic, and societal issues concerning the use of animals in toxicity studies (summarized under the 3Rs reduce, replace, refine), it is clear that the use of 3D cell culture and organoids will increase. The work carried out, to date, using proteomics to study organoids has been reviewed recently by Gonneaud et al. [32].

Table 4.4 Protocol for the preparation of liver microtissues for proteomic analysis by nLC-MS/MS

Cultivate microtissues in GravityTRAP™ plates with 70 μL 3D InSight™ Human Liver Maintenance Medium (InSphero AG) per well. Treat at day 5 with 0, 1.5, 4.5, 13.7, 41.2, 123, 5, 370.4, 1111.1, 3333.3, and 10,000 μM APAP (acetyl-para-aminophenol (paracetamol)) (dissolved in the medium) for 3 days without redosing
Measure biological triplicates of each concentration with the CellTiter-Glo® ATP-assay (Promega)
Pool 12 single microtissues from each condition in an Eppendorf tube
Spin at 200 g for 5 min at room temperature and then wash twice with PBS (sodium perborate) with the same spinning regime between washes
Lysis should be carried out in 20 μL of 10 M urea, 0.1 M ammonium bicarbonate, and 0.1% RapiGest, by sonication for 3 min followed by centrifugation at 16,000 g for 2 min at 18 °C
Reduce the lysate with 5 mM tris(2-carboxyethyl)phosphine for 1 h at 37 °C followed by alkylation with 25 mM iodoacetamide for 20 min at 21 °C
Dilute to 2 M urea and digest with trypsin at a ratio 1:100 (enzyme to protein) at 37 °C for 15 h
The samples should now be spun at 20,000 g at 4 °C for 10 min
Peptides can now be desalted using C18 MacroSpin columns
Resuspend after drying, in 1% ACN and 0.1% formic acid

Adapted from Bruderer et al. [31]

4.2.2 Skin Toxicoproteomics

The skin is the largest organ of the human body. In addition to its major functions, such as the prevention of desiccation and provision of protection against environmental hazards such as bacteria, chemicals, and UV radiation, it is a metabolizing organ and a route of excretion as well as aiding in the maintenance of body temperature (via the presence of sweat ducts) [33]. While proteomic analysis of skin may be carried out for a number of reasons, the study of skin diseases has accounted for the majority of the work that has been reported to date (Table 4.5).

Studies have also been carried out on the presence and effects of metabolizing enzymes [50] and wound healing (including the effect of blast, infection, and wound healing treatments) [51, 52, 61].

The paper published by Van Eijl et al. [50] on metabolizing enzymes in 2012 provides an excellent general methodology for skin proteomics. Here the object of the experiments conducted was to quantitatively compare the levels of metabolizing enzymes in human skin and the 3D reconstructed skin models Epiderm-200 (MatTek Corporation, Ashland, MA, USA), RHE (SkinEthic Laboratories, Lyon, France), and EpiSkin (SkinEthic Laboratories). The strategy adopted was the pre-fractionation of the proteins extracted from cystolic and microsomal fractions of samples using the GeLC-MS approach [62]. In this approach prior to LC/MS/MS analysis, the protein extract is separated using two-dimensional (2D) SDS-polyacrylamide gel electrophoresis (PAGE). The gel is not used to isolate individual proteins but rather as simply a way of subdividing the complex extract into

Table 4.5 Applications of proteomics in skin research

Application	Techniques used	Reference
(i) Disease studies		
Atopic dermatitis	MRM	[34]
Melanoma	MSI, nLC/MS/MS, DIA	[35–37]
Contact dermatitis	Review [16], DIGE (difference gel electrophoresis), nLC/MS/MS	[16, 38, 39]
Cutaneous T-cell lymphoma	Review	[40]
Psoriasis	nLC/MS/MS, DIA	[41–43]
Psoriatic arthritis	SCX (strong cation exchange), nLC/MS/MS, DIA, MRM	[42, 44]
(ii) Aging	DIGE, MALDI-MS	[45]
(iii) Radiation exposure		
Ionizing radiation	nLC/MS/MS	[46]
UV	nLC/MS/MS LC-MALDI-MS	[47–49]
(iv) Skin metabolism	GeLC-MS, DIA, MRM	[50]
(v) Wound healing	MSI, GeLC-MS, DIA	[51–56]
(vi) Skin irritation	GeLC-MS, ITRAQ	[57, 58]
(vii) Skin sensitization (mostly skin cells)	MSI	[38, 39, 43, 59, 60]

10–20 fractions by physically cutting the gel lane into equal length strips prior to enzymatic digest. For the preparation of human skin for analysis, the overall strategy adopted is summarized in Table 4.6.

In skin toxicoproteomics, a lot of effort has gone into the study of proteomic responses to irritants and sensitizers, and this will be the focus of the following section. Skin irritants are defined as substances that lead to the production of reversible damage to the skin following application for up to 4 h. In contrast skin sensitizers are substances that can lead to an allergic response following contact with the skin, such a response is termed allergic contact dermatitis (ACD) in humans. Sensitization evolves in two phases: The first phase is the induction of specialized immunological memory in an individual following exposure to an allergen. The second phase is elicitation, i.e., the production of a cell-mediated allergic response by exposure of a sensitized individual to the same allergen [https://eurl-ecvam.jrc.ec.europa.eu/validation-regulatory-acceptance/topical-toxicity/skin-sensitisation]. A major research effort into the study of skin irritation and sensitization was stimulated by European legislation, Directive 76/768 EEC which detailed with the upcoming prohibition of animal testing in the cosmetic industry.

An important outcome of this work and of relevance here was the development of the peptide reactivity assay [63]. This assay makes use of the concept of a molecular initiating event (MIE) [30] to assay a test substance for its potential to be a skin sensitizer based on its binding to model peptides containing an appropriate neutrophile.

The application of proteomics in the study of chemical-mediated allergic contact dermatitis has been reviewed recently by Höper et al. [38]. As is highlighted in this

Table 4.6 Protocol for the preparation of human skin for proteomic analysis of proteins involved in xenobiotic metabolism by GeLC-MS/MS

Skin samples (obtained with appropriate ethical consent) should be collected immediately following surgery, chilled on ice during transportation to the laboratory, and then stored at −80 °C until processed
Remove subcutaneous tissue carefully and cut skin samples into small pieces, and then homogenize in 1.5 volumes of ice-cold 250 mM potassium phosphate buffer, pH 7.25 containing 150 mM KCl and 1 mM EDTA
Prepare microsomal and cytosolic fractions by differential centrifugation and store at −80 °C until required
Separate proteins in the microsomal and cytosol fractions on 10% NuPAGE Novex bis-tris gels (Invitrogen Ltd., Paisley, UK)
Stain with Instant-BlueH; each sample-containing lane should then be manually cut into a series of 20 regions based on the position of molecular weight markers (SeaBlue MarkerH, Invitrogen Ltd., Paisley, UK) and the distribution of proteins observed in the gel
Digest each gel piece with trypsin, extract peptides and dry
In this study, dried samples were reconstituted and injected onto a reverse phase column and the eluted peptides analyzed by liquid chromatography tandem mass spectrometry online using an Agilent 1200 LC series (Agilent Technologies UK Ltd., Berkshire, UK) and a Thermo LTQ linear ion trap MS (Thermo Scientific, Hemel Hempstead, UK) as described previously
In this study, data analysis was restricted to those proteins with a putative role in xenobiotic metabolism, and if this is required, these can be selected in an automated fashion using PROTSIFT (https://github.com/jcupitt/protsift)

Adapted from Van Eijl et al [50]

review, the complexity of whole tissue extracts has led much proteomics studies of skin sensitization to use specific cell lines rather than whole tissues [38].

Proteomics studies of whole skin for the study of skin irritation have however been conducted [57, 58]. In 2014, Parkinson et al. carried out a comprehensive proteomics study of the response of human skin to exposure to the known chemical irritant sodium dodecyl sulfate (SDS). Of interest here is that studies were carried out both on ex vivo biopsy samples and on biopsy samples which were treated in vivo on the skin of healthy volunteers. In this study the samples were analyzed qualitatively using a GeLC-MS very similar to that described above, but this was followed up with a quantitative comparison between treated and untreated samples using the isobaric tag for relative and absolute quantitation (ITRAQ) technique [64]. ITRAQ is a chemical derivatization strategy that produces MS/MS "reporter" ions in the low m/z (m/z 114–121) area of the mass spectrum from a peptide. This area of an MS/MS spectrum obtained from a peptide has minimal background, and hence the relative areas of these peaks can be used for relative quantification since they correspond to the proportions of the labelled peptides. In this study the authors, probably correctly, claim to have produced "the most comprehensive qualitative and quantitative data set of skin proteome to date."

Mass spectrometry imaging (MSI) has also been applied to toxicoproteomics. Hart et al. [59] utilized MALD-MSI for the study of ex vivo human skin in a search for biomarkers of skin sensitization, and protocols for the preparation of skin for

Table 4.7 Protocol for the preparation of human skin for direct protein analysis by MALDI-MSP and MALDI-MSI

Sectioned frozen tissue specimens were at 12 μm in a cryostat, transfer to ITO conductive glass slides, and wash with isopropanol (2 × 30 seconds) to remove salts, lipids, and contaminants that can cause signal suppression
Robotically spot MALDI matrix, sinapinic acid onto regions of interest (ROI) within the tissue to be profiled for proteins
Identify the wound bed (WB) and adjacent dermis (AD) ROIs via histological evaluation of H&E-stained sections
Acquire and export MALDI-MS tissue profiles from selected wound bed (burn in this study) samples and normal skin (controls), baseline correct, and normalize to the total ion current (to generate spectra to be used in statistical analysis)
For imaging mass spectrometry analyses, matrix sublimation followed by a quick (4 min, 85 °C) rehydration and recrystallization step can be used to prepare the samples

Adapted from Taverna et al. [54]

imaging have been published [65–67]. Wound healing has been the subject of a number of MSI publications [35, 38, 54, 68, 69]. Taking as illustrative of these the 2016 study of burn wounds by Taverna et al. [54], the methodology used can be summarized as Table 4.7 shows, and it is a typical methodology for intact protein imaging in skin.

The study of wound healing is another area where the use of 3D cellular models, "living skin equivalents" (LSE) seems set to help to progress the 3Rs agenda [70]. While there are no reports of proteomics studies of wound healing in 3D models to date, the metabolomics work reported by Lewis et al. [70] indicates that this would fruitful area for further research.

4.3   Proteomics Goes Forensic: Sample Treatment for Blood and Fingermarks Proteomics

It is only in the recent years that proteomics has been employed to answer questions of forensic science nature. Perhaps this is due to the fact that the vast majority of information sought in forensic inquiries derive from the detection and identification of small molecules (drugs of abuse, medications, metabolites, alcohol) or from DNA recovery and analysis. However, as it will be discussed further in this section, proteins can also be an important source of forensic intelligence and this justifies the development and application of sample treatment strategies for proteomic analysis of some forensic specimens.

Although there is a wide range of forensic traces that can be retrieved at crime scenes, fingermarks are the most common and, at least in the UK, 2/3 of suspect identifications are still made through fingerprinting[1] despite the advent of DNA (Mr Neil Denison, Head of Identification Services, Yorkshire and Humber region, UK, personal

---

[1] The process of matching a crime scene fingermark to a fingerprint record held in National Databases.

communication). While fingermarks are not strictly considered "tissue" by the Human Tissue Act, this section will focus on this specimen because they contain residual cells and DNA, as well as because of the pivotal role often played during investigations and judicial debates. In particular, sample treatment strategies will be discussed in order to obtain additional forensic intelligence from this type of evidence.

While fingermarks are common to virtually every crime scene, blood is the most encountered type of evidence at a scene of a violent crime. Its detection can be paramount to reconstruct the dynamics of the crime and a timeline of the events as well as proving/disproving defendant statements and "legitimate access." For this reason, this section of the chapter will also include a discussion of proteomic strategies applied to blood with the aim to provide reliable and robust detection of this biofluid and contribute to a correct course of justice.

In particular, sample treatment strategies will be discussed which do not include the use of liquid chromatography prior to the mass spectrometric analysis.

The hyphenation with liquid chromatography, typically with electrospray mass spectrometry, ensures a much higher yield of protein identification, enabling a much more efficient tandem mass spectrometry to be applied for confirmation of identification. However, this technique bears the disadvantage to be more complex, time-consuming, and labor-intensive. For this reason, in situ proteomics and sample treatment for subsequent analysis by matrix-assisted laser desorption/ionization mass spectrometry (MALDI-MS) will be covered instead as it offers rapid analysis and simpler sample preparation.

4.3.1 In Situ Proteomic Strategies for Fingermark Evidence

Fingermarks are the result of the transfer of sweat and contaminants on the fingertip to a surface upon contact. Despite the advent of DNA, they are still a very important means of biometric identification in investigations leading to an arrest and/or a conviction. This type of evidence has been significantly investigated at a molecular level since 2008 thus creating a line of analysis parallel to conventional (physical) police examination based on the ridge detail and local characteristics (*minutiae*) making every fingermark unique.

Ifa et al. were the first scientists to report the use of a mass spectrometry technique, namely, desorption electrospray ionization mass spectrometry, to detect and map the molecular content of a fingermark directly onto the identifying ridge details [71]. Rowell's and Francese's groups followed shortly after with the development of surface-assisted laser desorption/ionization mass spectrometry (SALDI-MS) and MALDI-MSP/MALDI-MSI, respectively [72, 73]. However, out of the three techniques, MALDI-MS has been significantly the most extensively published [74] and the only one that has been included in the *Fingermark Visualisation Manual* edited by the Home Office UK [75] (as Category C, Technology Readiness level 3). Through the application of MALDI, mainly, and other mass spectrometric techniques, a variety of classes of molecules have been investigated within this specimen, from lipids to amino acids and inorganic compounds [76–79], ingested drugs or medications (and metabolites) [80, 81], as well as external contaminants such as explosives [82, 83], condom lubricants [84, 85], blood [86–88], and contact drugs and metabolites [89]. The detection and mapping of these substances may provide intelligence around the suspect, their lifestyle and possibly activities connected to the circumstances of the crime as well as crime dynamics.

The first example of peptide and protein studies in fingermarks was provided by the work of Ferguson et al. in 2012 [90]. Here the authors exploited the presence of characteristic peptides and proteins normally present in sweat to establish the sex of the donor on the basis of the differential profiles exhibited in men and women. The accuracy of prediction was 85% using a cohort of 80 donors excluding donors over 40 years old, smokers and donors taking of drugs/medications 2 weeks prior to the study. Partial least square discriminant analysis was employed with the discriminating species having a molecular weight in the range between 2000 and 9000 Da. The only sample treatment applied to the ungroomed marks, prepared as previously described [73], was the spotting of the MALDI matrix to enable MALDI-MSP analysis. The most efficient MALDI matrix composition was 5 mg/mL CHCA (alpha-cyano 4 hydroxycinnamic acid) in 25:25:50 acetonitrile-ethanol-0.5% TFA_{aq}. While promising and proving that the method was sensitive enough to detect material from a depleted specimen (ungroomed marks), the accuracy of prediction must be improved, and "natural fingermarks" would need to be investigated in a study with no exclusion criteria. This work is in progress in our laboratories and will give further insights into feasibility of application. The discriminating peptides and small proteins mostly are antimicrobial species (such as dermicidin and cathelicidin); they are produced in sweat and present on skin as a response to a variety of pathogenic microorganisms' defense. These species had already been detected in prior studies using either surface-enhanced laser desorption/ionization mass spectrometry (SELDI MS) or MALDI-MS, though sweat and not fingermarks, was the investigated specimen [91–93].

The detection of such species directly in fingermarks and the prospect of contributing to important forensic intelligence were the basis for subsequent development work involving in situ "tissue" proteomics. In particular, the development of such methods was aimed to: (i) confirm the identity of the discriminating species and (ii) possibly enhance the prediction of accuracy by using smaller m/z peptide/protein-deriving peptide sex biomarkers. Additionally, a thorough inspection of the literature indicated that many of the peptide/protein species that are possible to detect in fingermarks, including psoriasin and dermicidin (DCD)-derived peptides, may also act as biomarkers of breast cancer and other tumors [94–96]. This occurrence suggests the possibility to both provide a higher level of forensic intelligence (sex and possible disease detection) and of a cross-over with the biomedical world in the field of diagnostics. For these reasons, the development and optimization of proteomic strategies to detect peptide and protein species from fingermarks were pursued by Francese's group with great interest.

The only prior example of proteomic strategy for the identification of antimicrobial proteins was published by Baechle et al. in 2006 [93]. Here sweat, and not fingermarks, was separated on SDS page and the bands of interest excited and digested in situ using trypsin as proteolytic enzyme. It is important to note that no surfactant was employed within the trypsin solution.

It has been common knowledge for many years now that for in-gel, in-solution, and on-tissue digestions, the use of detergents in low concentrations greatly reduce steric hindrance enabling trypsin to cleave proteins (especially hydrophobic proteins) at many more specific cleavage sites than when a detergent is not employed [97–99]. For this reason Patel et al. [100] focused on the research and optimization of the most efficient detergent within trypsin digestions, particularly of fingermarks. The yield in the peptide ion population and ion abundance generated by the different detergents and detergent concentrations were assessed using a systematic approach and a quantitative evaluation. Ungroomed fingermarks were employed to simplify sample preparation as peptide ion suppression is avoided through the prior depletion of lipids within this specimen. However, from our unpublished studies, if natural fingermarks are employed, washes with denatured ethanol and chloroform in separate instances are recommended.

Firstly, a range of detergents was chosen for evaluation [100]. n-Octyl ß-D-glucopyranoside (OcGlu) was primarily selected as it was at the time, at large, the most used detergent for trypsin digestion. Two additional nonionic surfactants were also selected, namely, n-decyl ß-D-maltoside (DDM) and N-octanoyl-N-methylglucamine (MEGA-8) (and combinations thereof). Additionally, the anionic surfactant sodium 3-[(2-methyl-2-undecyl-1,3-dioxolan-4-yl)methoxyl]-1-propanesulfonate, marketed under the trade name of RapiGest SF, was also trialled; this detergent had been already reported for in-solution digestions [101, 102]. The structures of the investigated surfactants are shown in Fig. 4.6.

A number of variables were investigated concurring to the determination of the most efficient detergent, namely, (i) the mode of application of trypsin, (ii) the concentration of the detergent (optimized according to (i)), and (iii) the combination of multiple detergents to encompass the different unfolding capabilities in relation to different proteins in a complex mixture.

With respect to (i), the mode of deposition, namely, by spotting or homogeneously spraying the enzyme, is dictated by the type of desired outcome; if the purpose is to detect the proteome in discrete areas of the marks, then spotting is the method of choice. If the purpose is to detect and map peptides onto the fingermark ridges via MALDI-MSI, then trypsin must be sprayed, although in the work by Patel et al., no fingermark imaging analysis was eventually attempted. In both deposition modes, the trypsin was used at a concentration of 20 µg/mL containing either 10 mM OcGlu, OcThio, DDM, MEGA-8, or RapiGest SF, in concentrations varying from 0.5 to 2% w/v. When these solutions were spotted, 0.5 µL droplets were manually deposited onto the fingermarks. When these solutions were sprayed onto fingermarks using the SunCollect pneumatic sprayer, seven layers of trypsin were deposited at a flow rate of 1.5 µL/min.

In another experiment, 10 mM of each of three detergents, namely, OcGlu, MEGA-8, and DDM, were mixed in the ratios of 1:1:1, 1:2:1, and 1:1:2, and 0.5% w/v of this detergent mixture was finally added to 20 µg/mL trypsin solution.

The detailed overall conditions of proteolysis and further preparation for MALDI-MSP analysis are summarized in Table 4.8.

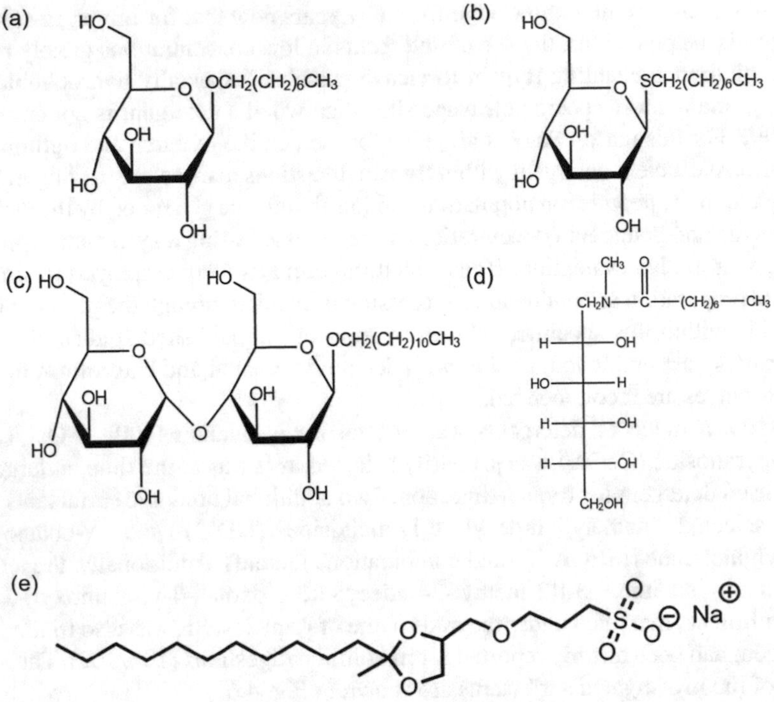

Fig. 4.6 Chemical structures of nonionic surfactants: (**a**) n-octyl ß-D-glucopyranoside (OcGlu), (**b**) n-octyl 1-thio-ß-D-glucopyranoside (OcThio), (**c**) n-decyl ß-D-maltoside (DDM), (**d**) n-octanoyl-N-methylglucamine (MEGA-8), and the anionic surfactant, (**e**) sodium 3-[(2-methyl-2-undecyl-1,3-dioxolan-4-yl)methoxyl]-1-propanesulfonate (RapiGest SF). (Reproduced and adapted (panel **e**) from Patel et al. [100] under the constraints of a Creative Commons Attribution 4.0 International License (https://creativecommons.org/licenses/by/4.0/legalcode))

Data Evaluation A manual inspection of all the spectra, produced under the different conditions reported above, enabled preliminary observations on the efficiency of the different detergents used in different concentrations. However, to enable a more objective and quantitative evaluation, following the analysis of the in silico digest of all the protein targets, ion signals were considered as "peptides" only if their *m/z* fell within the fractional range 0.4–0.8 (in line with what the proteomic community considers the fractional range of peptides to fall within). The *m/z* and intensity of these ion signals were used to enable a uniform relative comparison between performances of the different detergents within the two deposition scenarios.

As an example, Fig. 4.7 reports the mass spectra obtained upon trypsin spotting and proteolysis incorporating separately the four nonionic detergents at a given concentration (0.5%). The figure also shows the number of peptides within the fractional range 0.4–0.8 per given detergent (with range). With respect to the evaluation of the ion abundance, a reproducible ion matrix peak was selected and the ratio between the sum of the peptide intensity and this matrix peak was plotted per given

Table 4.8 Summary of the protocols employed for trypsin proteolysis of fingermarks as reported by Patel et al. [100]

	Detergent employed	Detergent conc. added to trypsin solution (% w/v)	Trypsin solution, application and incubation conditions	Matrix deposition
Spotting deposition method	OcGlu	0.5	20 µg/mL in 50 mM ammonium bicarbonate, pH 8 Upon detergent addition 0.5 µL manually spotted Incubation in a parafilm-covered jar containing 50:50 H_2O:MeOH for 3 h at 37 °C (5% CO_2)	Spotting of 0.5 µL of 10 mg/mL CHCA in 50:50 ACN:0.5% TFA_{aq} containing equimolar amounts of aniline to CHCA
	OcThio	0.5		
	DDM	0.5		
	MEGA-8	0.5		
Spraying deposition method	OcGlu	0.5	20 µg/mL in 50 mM ammonium bicarbonate, pH 8 Upon detergent addition seven layers of trypsin were sprayed at a flow rate of 1.5 µL/min Incubation in a parafilm-covered jar containing 50:50 H_2O:MeOH for 3 h at 37 °C (5% CO_2)	5 layers of 5 mg/mL CHCA in 50:50 ACN:0.5% TFA_{aq} containing equimolar amounts of aniline, sprayed at a flow rate of 1.5 µL/min
		1		
		2		
	DDM	0.5		
		1		
		2		
	MEGA-8	0.5		
		1		
		2		
	OcGlu:MEGA-8:DDM 10 mM each in ratios: 1:1:1, 1:2:1, 1:1:2	0.5		
	RapiGest SF	0.1		

detergent (with range). This type of data evaluation enabled to conclude that in spotting experiments repeated in triplicates, all the detergents yielded a comparable number of peptides which is to be expected given the fairly similar detergent structures. However, OcGlu, followed by OcThio, showed the largest range (lower reproducibility) as opposed to DDM and MEGA-8 showing the highest reproducibility in the yield of the ion population. However, when the ion abundance was examined, while OcThio produced the most intense peptides, it also showed the lowest reproducibility and was therefore not taken forward in subsequent experiments in which the detergent containing trypsin solution was sprayed rather than spotted.

In the "spraying" experiment, different concentrations of the detergents and their performance were evaluated in the same way. Results are summarized in Fig. 4.8 showing that when used at a 2% concentration, MEGA-8 generated the most reproducible peptide numbers out of the three detergents tested together, with highest number of peptides.

This observation indicated MEGA-8 at 2% as the highest performing detergent in spray deposition conditions, despite, in these conditions, for all the other deter-

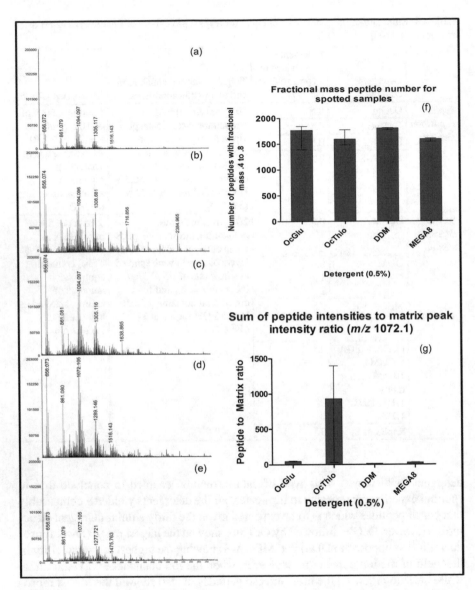

Fig. 4.7 MALDI-MS peptide profiles from in situ digests of ungroomed fingermarks spotted with 20 µg/mL trypsin alone in 50 mM ammonium bicarbonate pH 8.04 (**a**), or containing 0.5% concentration of the detergents (**b**) OcGlu, (**c**) OcThio, (**d**) DDM, or (**e**) MEGA-8. Column graph showing the number of peptides with fractional mass between 0.4 and 0.8 (**f**) and corresponding peptide/matrix intensity ratio for each detergent (**g**). The consistently detected matrix peak at *m/z* 1072.1047 ([CHCA − 4H + 4Na + 1 K]+) was selected for the calculation of this ratio. (Reproduced from Patel et al. [100] under the constraints of a Creative Commons Attribution 4.0 International License (https://creativecommons.org/licenses/by/4.0/legalcode))

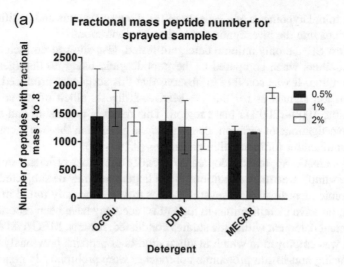

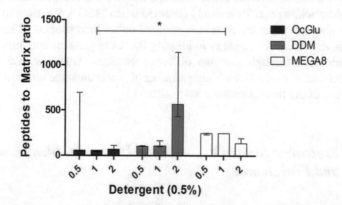

Fig. 4.8 Column graph of the number of peptides with fractional mass between 0.4 and 0.8 for OcGlu, DDM, and MEGA-8 at three concentrations (**a**). Graphical representation of peptide intensity/matrix intensity ratios for each detergent at three concentrations. Statistical analysis calculated a significant increase between MEGA-8 and OcGlu at 1%, as denoted by the asterisk. (Reproduced and adapted from Patel et al. [100] under the constraints of a Creative Commons Attribution 4.0 International License (https://creativecommons.org/licenses/by/4.0/legalcode))

gents, 0.5% concentration yields the highest peptide intensity/matrix intensity ratio. It is speculated that for MEGA-8, a better ratio detergent/endogenous proteins and/ or trypsin/endogenous proteins occurs in spray conditions over spotting experiments.

When the three detergents were used in mixture in different ratios, the main observation for the optimal 0.5% concentration, was an increase in the ion population beyond the *m/z* 1800 yielded by the individual detergents. This occurrence does verify the initial hypothesis of a synergy of the three surfactants and justifies further investigations into the investigation of higher concentrations.

RapiGest SF, the only anionic detergent tested, also showed an increase in the peptide numbers when compared to the peptide yield using no detergents in the trypsin solution. It was possible to observe that this surfactant produced a similar number of ion signals as for the 2% MEGA-8 digests albeit of lower intensity, except in the 1000–1200 Da mass region. This result is promising and worthy of further investigations of different concentrations other than the one recommended by the manufacturer for in-solution digestions (0.1% w/v).

The systematic approach adopted to identify the most efficient surfactant(s) within the sample treatment conditions that a fingermark may be subjected to, led to the proteomic identification of several species that are normally found in sweat. In particular, putative identifications in high mass accuracy have been reported for the most efficient detergent within the studies conducted, namely, MEGA-8. A targeted approach was employed in which in silico digests of proteins previously identified in sweat using non-in situ proteomic approaches were preliminarily generated and then surveyed [103, 104]. Table 4.9 shows a comparison of the proteins identified when MEGA-8 was spotted or sprayed, respectively. It is interesting to note that the putatively identified proteins differ between the two sample treatment approaches and that dermicidin peptides are only detected when MEGA-8 is used in the higher concentration and when the trypsin is sprayed rather than spotted. Clearly, further method development is necessary to identify the set of proteins encompassing both subsets obtained through the two different methods. However, the spray-coat method does seem to identify a higher number of low abundance proteins in contrast with the use of the trypsin spotting approach.

4.3.2 Proteomic Strategies for Blood Detection, Identification, and Provenance

Background – Blood is a very common biofluid found at the scene of violent crimes, and its reliable detection is extremely important to reconstruct events that have taken place at the crime scene as well as for the detection of forensically interesting substances potentially carried by blood. This biofluid can be found in stains and in association with fingermarks. However, its detection and confirmation is not as obvious as it may seem even when a red coloration is observed; furthermore blood can be present only in invisible traces as a result of an attempted "clean-up" of the scene by the perpetrator or because originally already present in minute amounts. It is then down to the sensitivity of the technique applied whether or not blood evidence can be recovered and confirmed. For stains it is somewhat easier than fingermarks as commonly heme-reactive reagents like luminol can be used in generous amounts by spraying the environment and observing fluorescence in the dark (should blood be present). In fact, luminol would also show an attempt to clean up the scene. However, the use of this reagent would destroy the ridge detail of any blood mark. In addition,

Table 4.9 Peptide mass fingerprinting and putative protein identifications from in situ fingermark digests performed by spotting or spraying a trypsin solution containing 0.5% MEGA-8 or 2% MEGA-8, respectively, as a detergent

Protein	0.2% MEGA-8 spotted	Peptide *m/z* and sequence	2% MEGA-8 sprayed	Peptide *m/z* and sequence
Adrenomedullin			√	1060.5605 SIGTFSDPCKDPTRITSPNDPCLTGK
Alpha-2-glycoprotein 1 zinc	√	2035.0434 IDVHWTRAGEVQEPELR		
Antibacterial protein LL-37134–170			√	1365.6610 WALSRGKR
Aspartate aminotransferase mitochondrial	√	2216.9593 NLFAFFDMAYQGFASGDGDK [met ox]		
Beta-defensin 103 precursor			√	703.3621 EEQIGK 933.4896 CAVLSCLPK
Calmodulin-like protein 3	√	2117.0760 ELGTVMRSLGQNPTEAELR [met ox]		
Corneodesmosin	√	2750.2882 SIGTFSDPCKDPTRITSPNDPCLTGK		
Dermicidin			√	676.3829 SSLLEK 725.3932 GAVHDVK 1128.5365 ENAGEDPGLAR 1459.7622 LGKDAVEDLESVGK 1466.7872 GAVHDVKDVLDSVL

(continued)

Table 4.9 (continued)

	0.2% MEGA-8 spotted	Peptide *m/z* and sequence	2% MEGA-8 sprayed	Peptide *m/z* and sequence
Filamin B	√	2064.0684 SPFEVSVDKAQGDASKVTAK		
Human serum albumin	√	1898.9799 RHPYFYAPELLFFAK 2346.9887 TYETTLEKCCAAADPHECYAK		
Kallikrein-11	√	2102.9956 GFECKPHSQPWQAALFEK 2403.0369 CENAYPGNITDTMVCASVQEGGK [met ox]		
Keratin 1B			√	967.4723 DVDAAYVSK
Keratin type I	√	2187.0196 SDLEMQYETLQEELMALK [met ox]	√	1060.5605 TLLDIDNTR 1323.6725 IKFEMEQNLR [met ox]
Psoriasin	√	2750.2882 ENFPNFLSACDKKGTNYLADVFEK	√	1384.7194 KGTNYLADVFEK

all the blood enhancement techniques (BET), whether targeting stains of blood fingermarks, are only presumptive, that is, they provide an indication that blood may be present or not but they are not confirmatory and they may lead to false positives. These methods have been extensively reviewed and all were reported to exhibit a lack of specificity [105]; even heme-reactive compounds, the most specific class of blood reagents, may give false positives as, for instance, horseradish, leather, and other extracts from plant material show the same peroxidase activity exhibited by heme in human blood. As a further example, protein dyes are used as BET, working on the attraction of the dyes' negatively charged sulfonate group to the cationic group of proteins. While it is true that proteins are present in great abundance in blood, since they are also present in other biofluids such as semen and saliva, these agents could well lead to false positives.

It is therefore paramount to develop a methodology enabling the specific (and therefore reliable) determination of the presence of blood. Though heme and hemoglobin (respectively) had already been analyzed and detected by mass spectrometry, Francese's group was the first to use MALDI-MSP and MALDI-MSI for the intact detection of these molecules as markers of blood presence in a forensic context [86]. Furthermore, the presence of these molecules was visualized directly onto the identifying ridge detail of fingermarks (fresh and 7 day old) in non-enhanced and BET-enhanced marks (or previously BET-enhanced stains), thus providing the link between the biometric information and the circumstances of the crime. It was possible to calculate that, on the basis of average concentration of hemoglobin (and heme) in healthy patients, hemoglobin and heme could be detected down to 1000 times and 250 million times, respectively, lower than physiological concentration. A few more experiments were reported in this published work [86] showing the possibility to determine blood provenance by distinguishing between human, equine, and bovine intact hemoglobin exploiting the subtle differences in the amino acid sequence of the protein.

This methodology involved little sample treatment and exclusively related to the preparation for MALDI-MSP and MALDI-MSI analysis; for fingermarks, they were simply sprayed with a 5 mg/mL solution of α-cyano 4-hydroxycinnamic acid in 70/30 acetonitrile/TFAaq 0.5%, whereas stains were preliminarily tape lifted and spotted directly with the same matrix solution, albeit in concentration of 20 mg/mL ready for MALDI-MS analysis.

This work represented a step forward in terms of the possibility to provide investigations and judicial debates with more reliable evidence. In fact, the presence of blood is claimed no longer by inferring it from a colour change due to staining or catalytic reaction but based on the actual determination of the presence of characteristic blood molecules. However, despite the easy and quick sample treatment, the mass accuracy on large biomolecules such as hemoglobin is limited and may not permit adequate detection and blood provenance, especially with mixed blood sources. The mass accuracy that can be achieved on the protein-deriving peptides is much higher (few parts per million); hence a greater specificity (hence reliability of the information) can be achieved by using a bottom-up proteomic approach which would also enable the detection of additional blood-specific proteins.

When surveying the literature in 2015, there were already many reports attempting to map the proteome of plasma and serum, which were extensively reviewed by Liumbruno et al. [106], and none of the approaches had involved the direct application of MALDI-MS on enzymatically digested blood. This is understandable as in all of the previous reports the aim was to map the entirety of the blood proteome for medical and diagnostic purposes. All the proposed approaches consisted in the hyphenation of some kind of separation technique with mass spectrometry (with LC-ESI MS/MS being the most proficient); individual approaches yielded typically a few hundred proteins leading to more than 10,500 blood/serum proteins being identified in total so far. The most elegant approach yet was proposed by Martin et al. [107] by coupling liquid extraction surface analysis (LESA) with LC-ESI MS/MS; in their study dried blood spots were investigated and the approach yielded the identification of over 100 proteins in an untargeted approach. Here sample extraction from the DBS (Dried blood spot) analysis and digestion was performed robotically, and 0.1 μg/μL trypsin solution (not containing any surfactant) was employed to digest the blood proteins for 1 h prior to transfer to the HPLC (high-performance liquid chromatography) autosampler and analysis by LC-MS/MS.

However, in a forensic context, the reliable detection of a handful of blood-specific proteins via bottom-up proteomic approach, using MALDI-MS if possible, would be more than appropriate. Furthermore, in forensic science, provided that reliability of the evidence is not compromised, speed is paramount to investigations; the hyphenated methods reported can be very labor-intensive and time-consuming, especially since some of them employ preliminary purification to deplete blood of the most abundant proteins. The literature offers information on the most abundant and blood-specific proteins [108–111] which are expected to be detected and that can be used in a targeted proteomic approach. Hemoglobin, haptoglobin, fibrinogen, transferrin, α-1-antitrypsin, and α-2-macroglobulin, to name a few, are among the most abundant.

In 2017, an Australian research group published a report on the use of an in situ proteomic approach targeting hemoglobin and enabling the discrimination of human and animal blood through the detection of proteotypic peptides [112]. The work described both sample extraction and in situ sample treatments for bloodstains and blood marks and the possibility to retrieve MALDI molecular images of the Hb peptidic signatures for the different species. However, the image resolution and the localization of proteotypic peptides on the mark were poor due to the deposition method (spotting of a large volume of enzyme on the mark) and the low spatial resolution with which the data were acquired.

Prior to this work, in 2015, supported by the Home Office Centre for Applied Science and Technology (UK), Francese's group embarked in a research program for the development of a multi-informative and specific methodology for detection of blood in fingermarks and stains via MALDI-MS-based strategies. Differently from the 2017 report, the methodology targets multiple blood-specific proteins, thus increasing the overall method specificity. A method for a homogenous and fine deposition of the proteolytic enzyme has also been devised enabling high resolution mapping of blood signature onto the fingermark ridge details. Sample treatment protocols and the effect on the reliability, robustness, and versatility of the methodology will be discussed in the next section.

Sample Treatment for Proteomic Analysis of Blood Stains   Initial protocol design was optimized for bloodstains with the view to cover the analysis of this type of trace evidence but also to adapt the methodology for the nondestructive detection of blood in fingermarks. While direct in situ analysis can be performed on bloodstains, general analytical community findings are that a preliminary extraction followed by an in-solution digest typically yields a higher number of peptides. Furthermore, differently from fingermarks, swabbing is a much more commonly applied forensic practice for bloodstains. Because of the latter, and also due to the challenges connected with an efficient tape lift for older and/or BET-enhanced bloodstains, a sample treatment involving extraction of the blood trace was mainly developed for this type of blood evidence.

Depending on the state in which the evidence is available, two types of blood retrieval techniques and subsequent treatment were applied [87]:

(i). Pipetting 70 µL of 50% ACN_{aq} solution directly onto the dried blood regions; the extract was then transferred to an Eppendorf where 50/50 ACN/H_2O was added up to 1 mL in volume and subsequently sonicated. Forty microliters of 40 mM ammonium bicarbonate (AmBic) (pH 8) were then added to only 10 µL of the blood extracts.

(ii). Swabbing a small blood containing area (of a 9-year-old Acid Black 1 enhanced palm print) using a swab previously wetted with 70/30 ACN/H_2O; the tip of the swab was then cut and sonicated in 1 mL 70/30 ACN/H_2O (same solution employed to pipet and extract blood in case (i)) to release the proteins. Twenty microliters of the supernatant were then dried and redissolved in 20 µL of 50 mM AmBic (pH 8) in preparation for proteolysis. Many of the commonly used swabs contain PEG (polyethylene glycol) material. This occurrence was uncovered much later down the line of our research, and in order to avoid peptide ion suppression, polymer-free swabs only are now used in our labs.

(iii). Rolling silica-free aluminum slides containing dry blood into a glass vial and immersion in 1 mL 50% ACN solution and sonicated.

Table 4.10 summarizes sample treatment conditions prior to enzymatic digestion.

In terms of the actual enzymatic digestion, two approaches were initially compared and contrasted: (a) the classic in-solution digest and (b) the lab-on-plate approach.

In case (a), 9 µL of 20 µg/mL Trypsin Gold including 0.1% RapiGest™ SF were added to a 10 µL blood extract (i) or fresh blood. One hour of incubation time was sufficient to achieve most efficient proteolysis while minimizing trypsin autolysis. RapiGest™ SF was previously trialled in in situ digestion of fingermarks [100] and brain tissue sections indicating a viable alternative to the use of the classic detergent n-Octyl ß-D-glucopyranoside.

Table 4.10 Sample treatment for the recovery and extraction of blood prior to proteolysis

Mode of blood recovery	Extracting solution	Extracting conditions	Further treatment prior to proteolysis	Digestion method
(i) Pipetting extracting solution of blood containing surface	50% ACN$_{aq}$ solution	Addition of 50/50 ACN/ H$_2$O up to 1 mL to the transferred 70 µL extracting solution - ultrasonication for 10 min at 45 Hz frequency	Addition of 40 mM AmBic (pH 8) in a volume ratio 1/4 extract/AmBic solution (50 µL in total)	In-solution
(ii) Swabbing (with a swab prewetted with a 70% ACN$_{aq}$ solution)	70% ACN$_{aq}$ solution	Ultrasonication in 1 mL of 70% ACN$_{aq}$ solution for 10 min at 45 Hz frequency	50 mM AmBic (pH 8) 1/1 with the blood extract (40 µL in total)	In-solution Lab-on-plate
(iii) Immersion (of a rolled specimen containing blood)	50% ACN$_{aq}$ solution	Ultrasonication in 1 mL of 50% ACN$_{aq}$ solution for 10 min at 45 Hz frequency	---	Lab-on-plate
(iv) *Undeposited fresh blood*	NA	NA	NA	In-solution Lab-on-plate

Table 4.11 Proteolysis conditions for the in classic solution digest (a) and the lab-on-plate approach (b)

	In-solution digest	Lab-on-plate
Trypsin, trypsin concentration and additives	20 µg/mL *Trypsin Gold* including 0.1% RapiGest™ SF	6 mg/mL *Bovine Trypsin*
Volume of blood/extracted blood	10–20 µL	1 µL
Proteolysis conditions	37 °C and 5% CO$_2$	Room temperature (noncontrolled conditions)
Proteolysis time	60 min	5 min

In case (b), typically 1 µL (1/10 of the volume in case (a)) of blood extract (iii) or fresh blood was spotted on Vmh2-adsorbed enzyme MALDI plate wells which contained immobilized trypsin prepared as previously described [113]; the proteolysis was carried out for 5 min (1/12 of the time necessary for the optimized in-solution digest).

Vmh2 belongs to the class I hydrophobins. These proteins have demonstrated to homogeneously self-assemble on hydrophilic or hydrophobic surfaces [114] and to be capable of strongly binding enzymes such as trypsin (and other proteins) in their active form. In particular, the use of Vmh2 has been proposed as an easy and effective desalting protocol [115]; the method bears the advantages of a much shorter proteolysis time for protein in the nanofemtomolar range and a significant increase of the resulting peptides signal to noise (S/N). Table 4.11 reports trypsin digestion conditions using the classic in-solution and the lab-on-plate approaches.

In preparation for MALDI-MS and MS/MS analysis, a CHCA matrix in 50/50 ACN/0.5% TFA$_{aq}$ containing 4.8 µL aniline was either spotted directly on the Vmh2-coated trypsin immobilized MALDI plate well (lab-on-plate approach) or mixed 1/1 with the digested blood extract/fresh blood on the well of an untreated MALDI plate.

The two approaches enabled the confirmation of the blood presence through numerous blood peptide signatures (between 1 and 11 per protein) belonging to 10 proteins including hemoglobin alpha and beta chains, Complement C3, myoglobin, apolipoprotein A1, hemopexin, α-antitrypsin 1, serotransferrin, EBP (erythrocyte band protein) 4.2 and EBP 3, and α-2 macroglobulin. These blood-specific proteins were detected using peptide mass fingerprinting and predominantly exploiting the mass accuracy (expressed as relative error in ppm) as a match acceptance criterion, based on the mass accuracy of the mass spectrometer upon calibration prior to sample data acquisition; mass accuracy ranged between 0 and 11 pm for both the classic in-solution digest and the lab-on-plate approach. On a first observation, while the peptide intensities were higher using the former digestion method, the lab-on-plate approach provided generally better mass accuracy. However, it must be noted that higher peptide intensity was only observed in the case of human blood; for equine blood this occurrence was inverted with the lab-on-plate approach showing higher peptide intensity. This can be explained by the complexity of the sample; horse blood was in fact defibrinated, and therefore it is less complex than the whole human blood that was investigated in parallel.

Additionally, although both approaches yielded the detection of the same proteins, the classic in-solution digest generally yielded more peptide signatures per identified protein. While, on these bases, it may be obvious to choose the classic in-solution digest, the much reduced digestion time within the lab-on-plate approach (1/12) may outweigh the greater number of peptides detected with the in-solution method. This is something that is worthwhile to investigate in a validation study in parallel with LC-MS/MS analysis to verify robustness of the nonhyphenated methods.

Both approaches were also employed to determine the presence of blood in swabbed sample from a 9-year-old Acid Black 1 developed palm print present on a ceramic tile. From this old and already enhanced specimen, only 3 proteins were detected, namely, Hb (αHb and βHb peptides), EBP 4.2, and Complement C3, with a total of 12 peptides. However, Complement C3 (which is not highly specific to blood) was only detected through one peptide and only via the in-solution digest albeit with very high mass accuracy (1.3 ppm). Conversely, EBP 4.2 which is a highly blood-specific protein was detected by both approaches and through two different peptides (one per approach). Both peptides are proteotypic and indicate blood of human origin. From these analyses it was therefore clear that these methods may open a new avenue for the re-exhamination of cold cases. At this stage, the lab-on-plate approach would be recommended first due to its rapidity, though, if positive, a confirmatory test would be advised through the classic in-solution digest. In fact, in situ MS/MS analysis would be an even better direct confirmatory test. This type of analysis was attempted in this context and proved very useful not only to confirm the presence of blood but also to confirm the presence of blood of mixed provenance.

While the m/z assignment had already provided the indication that blood was originating from both human and equine source, ion mobility MS/MS allowed confirmation of the presence of equine blood through the αHb peptide at m/z 1499.7237.

These approaches were considered highly promising and further developed for blood in association with fingermarks.

Sample Treatment for Proteomic Analysis of Blood Marks As for bloodstains, BET currently used to visualize blood marks are largely presumptive. However, spectroscopic nondestructive techniques such as Raman [116–122] and hyperspectral imaging [123] have been reported for their ability to detect and map blood in a fingermark. While they might be in some cases more specific than BET, they all present other limitations including peak shifts as a function of laser power and blood age (Raman), unsuitability of red and dark substrates, and the requirement for a reference spectrum of a nonblood containing area (hyperspectral imaging), which, in a crime scene scenario, cannot be guaranteed to be available.

For blood marks, it would be highly beneficial to maintain the integrity of the evidence (ridge pattern) while proving the association of a fingermark with blood. If the integrity of the sample is preserved, a crucial associative link between the biometric information and the event of bloodshed can be established. Therefore, a technique capable of producing molecular images of a blood mark is desirable; while hyperspectral imaging can produce maps of analytical signatures, the above-reported limitations make the development of an alternative technique/method desirable. As previously mentioned, MALDI is capable of imaging molecular signatures (much more specific than analytical signatures) and an appropriate methodology was developed on the basis of the work by Bradshaw et al. and Patel et al. [86, 87] bridging the gap between the two studies.

While the adaptation of the lab-on-plate approach is certainly more challenging and is presently under investigation, the initial focus of subsequent work was the adaptation of the sample treatment developed for bloodstains in order to carry out an in situ (rather than in-solution) analysis of blood marks.

It is important to bear in mind that there are three types of association relating to three different dynamics at the scene: (a) blood mark, the fingertip was covered in blood and made contact with a clean surface; (b) mark in blood, the blood was present on a surface and a clean fingertip made contact with the biofluid containing surface; and (c) mark in coincidental association with blood, a clean fingertip made contact with a surface and the blood landed on it as a result of blood spatter. This type of information is extremely important for the building of a case. However, this section will not be covering such work in progress as this is being pursued with nonproteomics-based techniques and partly published [124]. The first step remains the demonstration that blood and no other biofluids or blood-resembling substances are associated with a mark.

In order to maintain intact the ridge pattern, the proteolytic enzyme must be sprayed over the blood mark in a manner that does not delocalize molecules or cause ridge merging. Not only will the conditions of a homogeneous spraying need to be devised (number of trypsin layers, speed of spraying, distance between the nozzle and sample to be sprayed), but it is paramount that the concentration of trypsin is also optimized. As it is well known, there is an optimal ratio between trypsin and the sample to digest (50:1 typically), above which, the mass spectrum may be very densely populated by trypsin autolysis ion signals and below which the proteolysis is incomplete/inefficient. Inevitably, this task can only start by using reference samples containing approximately the same amount of blood and spread as homogeneously as possible on the same surface area. Of course, crime scene marks in association with blood may be very varied in both the volume of blood present and in the differential distribution over the mark itself. With respect to the different volumes of blood, a study in progress is systematically investigating the yield of peptide blood signatures in relation to a fixed trypsin concentration and varied volumes of blood. This will establish validity of the use of a certain trypsin concentration for a range of defined blood volumes before concentration of the enzyme has to be raised in order to efficiently digest the next blood volumes range. At this stage, for real crime scenes, blood volumes will have to be correlated with visual appearance of the blood and the estimation of the volume will then correlate to the concentration of trypsin to employ. The choice of sample treatment conditions becomes more complex with differential volumes of blood present in different areas of the same mark, which is the norm in a forensic scenario. In these cases, if the aim is to provide molecular images of blood-specific proteins by visualizing them on the ridge pattern, it is suggested to privilege the choice of the trypsin concentration which is deemed to be appropriate for the volume of blood estimated to be present in the area where ridge pattern is shown rather than where blood pools are.

The optimization of the trypsin concentration for complex biological fluids has its limitations due to the fact that blood-specific proteins vary greatly in concentration or "accessibility" (e.g., EBP proteins which are erythrocyte membrane proteins). Therefore, it is to be expected that the combination of this type of limitation, deriving from varying blood volumes, and use of an in situ technique with no hyphenation would yield only a handful of blood-specific proteins.

In the work by Deininger et al. [88], following preliminary spot concentration tests, the concentration of (gold) trypsin was trialled on reference samples at values of 100, 150, 200, and 250 μg/mL (from 5 to 12.5 times higher than the classic in-solution digestion) incubating the sample for 3 h (3 times as much as the in-solution digest to aid trypsin accessibility for a non-in-solution sample), and the overall proteolysis conditions are reported in Table 4.12.

Table 4.12 Proteolysis conditions for in situ digest of a blood mark

	In-solution digest
Volume of blood/extracted blood	~100 to 200 μL
Trypsin, trypsin concentration and additives	100 or 150, 200 or 250 μg/mL *Trypsin Gold* in AMBIC 50 mM, pH 8, including 0.1% RapiGest™ SF
Spraying conditions on a SunCollect autosprayer	9 layers of trypsin at a flow rate 2 μL/min and a nitrogen pressure of 3 bar
Proteolysis conditions	Samples placed on polystyrene floats in a Coplin jar half-filled with 50:50 methanol:H_2O, sealed with parafilm; the jar's lid was wrapped in paper tissue. 37 °C and 5% CO_2
Proteolysis time	180 min

In particular, each blood mark was split in quarters and each quarter treated with a different trypsin concentration. Following on proteolysis, the blood marks were sprayed, using the automatic sprayer SunCollect, with five layers of 5 mg/mL CHCA in 70:30 ACN:0.2%TFA$_{aq}$, containing equimolar amounts of aniline to CHCA at a flow rate 2 μL/min and a nitrogen pressure of 3 bar.

The samples were then analyzed in MALDI-MSI mode on a Synapt G2 HDMS (Waters Corporation, Manchester, UK).

As a whole, the optimized analytical conditions have led to the detection and clear mapping of four blood proteins, namely, hemoglobin α-chain (αHb), hemopexin, serotransferrin, and Complement C (two peptides), with a relative error ranging between 3.2 and 18 ppm (Fig. 4.9).

A control experiment using the same donor's fingermark, which was blood-free, had been performed at a trypsin concentration yielding blood protein signatures in the blood marks. The absence of these signatures in the blood-free mark demonstrated that those ion signals did arise from blood presence in the blood mark. Although MS/MS could not be extensively applied due to the very low intensity of the relevant peptides, the application of ion mobility MS/MS did confirm the presence of both Hb α and β chains. Three out of four trialled trypsin concentrations, namely, 100, 150, and 200 μg/mL, yielded molecular images of blood peptide signatures, although the 150 μg/mL and 200 μg/mL concentrations seemed to perform better. The 250 μg/mL concentration of trypsin caused a delivery capillary blockage due to the high viscosity of the solution, and the corresponding quarter was therefore not appropriately trypsinized, thus yielding no molecular images.

Recently acquired data show that by using a larger internal diameter capillary (ID 100 μm), a trypsin concentration of 250 μg/mL could be delivered and that in fact was the most suitable concentration for the type of controlled samples used so far (Lisa Deininger, PhD Thesis, submitted January 2018).

The results from this study further demonstrated the possibility to contribute to forensic investigations involving the determination of the blood presence. However, the next step for blood mark imaging will be the application of the developed methodology to blood marks that have been lifted from surfaces, if the surface is not thin and flat enough to be inserted in the mass spectrometer. To date, the methodologies developed for bloodstains have been applied to an extract from a swab of 37-year-old blood mark on fabric enhanced using ninhydrin. MALDI-MS spectra show the presence of heme and peptides from both hemoglobin alpha and beta chains [125]. This occurrence showed the potential to image the whole mark by preliminarily digesting in situ and preparing it for a direct insertion in the mass spectrometer (when the surface of deposition is fabric). Though for very old marks, it would not be expected more than hemoglobin peptides, presence of Hb peptides would pave the way to direct an in situ blood fingermark analysis as well as demonstrate again the possibility to contribute to cold cases.

Currently, data from a validation study, applying the methods developed for bloodstains and blood marks to blind samples, are being evaluated. The samples

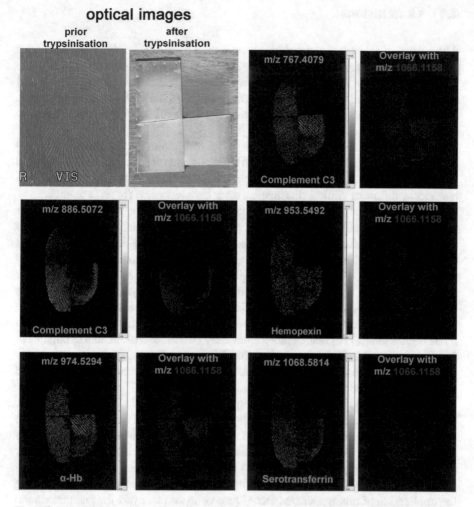

Fig. 4.9 MALDI-MSI of in situ proteolysis of a blood fingermark. The figure shows molecular images of blood-specific peptides generated by spraying trypsin in four different concentrations (100, 150, 200, and 250 μg/mL) on the blood mark using the SunCollect. The trypsin concentration of 250 μg/mL could not be delivered to due to the internal diameter of the delivery capillary. Each peptide image has also been overlaid with the matrix signal at *m/z* 1066.1158. The figure suggests that the best ridge reconstruction performance could be achieved using a trypsin concentration of/between 150 and 200 μg/mL. (Reprinted from Deininger et al. [88] with permission from John Wiley and Sons)

contain blood as well as other biological matrices (human biofluid and other matrices which may test positive for blood) and are both in a BET-enhanced and non-BET-enhanced form. The determination of both negative and false positives will enable a first strong evaluation of the applicability of this method in an operational context.

4.4   Conclusions

Although a widely practiced technique, on-tissue digestion imaging still has many potential courses for optimization to meet the demands of cutting-edge proteomics. The requirement for the best buffer, enzyme, solvent, and matrix could indeed reveal numerous concealed biological molecules, essential in the understanding of cancer, response to drugs, and also the recovery of useful intelligence in forensic investigations.

With continuous advancements in tissue preparations and treatments for proteomics involving techniques such as MALDI-MS and LESA together with the gold standard LC-MS/MS, these methods have the potential to forge a place in the systematic workflow of clinical diagnostics, drug discovery and forensic science.

In cancer proteomics and in toxicoproteomics, within the assessment of proteins that are pathologically relevant and therapeutic targets, respectively, it is important to appreciate how drug-treated tissues potentially affect dose-response relationship data. The effects of post-translational modifications could mask a biomarker during a time course, i.e., phosphorylation of peptides due to having a serine/threonine/tyrosine present in the sequence or having undergone methylation/acetylation, the latter being other bounds to conquer during pretreatment protocol design.

Toxicoproteomics is a field that has received limited attention to date and indeed it has been argued [126] that metabolomics may be a more fruitful area to pursue for this application. However, the studies that have been conducted to date have produced useful data that particularly in case of DILI and skin sensitization has yielded new biomarkers and a test protocol, respectively.

In forensic science and with respect to evidence such as fingermarks and blood, the sample preparation challenges to conquer are the minute amount of proteins in the former and the high dynamic protein range in the latter, where extremely abundant proteins, such as albumin and hemoglobin, may mask the presence of lower abundance proteins which are specific to blood. For both these specimens, the ultimate challenge remains the design of optimized sample treatment protocols enabling the application of nonhyphenated techniques or in situ analysis for fast turnaround of intelligence.

References

1. Stingl C, Soderquist M, Karlsson O, Boren M, Luider TM (2014) Uncovering effects of ex vivo protease activity during proteomics and peptidomics sample extraction in rat brain tissue by oxygen-18 labeling. J Proteome Res 13:2807–2817
2. Schober Y, Guenther S, Spengler B, Rompp A (2012) High-resolution matrix-assisted laser desorption/ionization imaging of tryptic peptides from tissue. Rapid Commun Mass Spectrom 26:1141–1146
3. Cole LM, Djidja MC, Bluff J, Claude E, Carolan VA, Paley M et al (2011) Investigation of protein induction in tumour vascular targeted strategies by MALDI MSI. Methods 54:442–453

4. Stauber J, MacAleese L, Franck J, Claude E, Snel M, Kaletas BK et al (2010) On-tissue protein identification and imaging by MALDI-ion mobility mass spectrometry. J Am Soc Mass Spectrom 21:338–347

5. Mao X, He J, Li T, Lu Z, Sun J, Meng Y et al (2016) Application of imaging mass spectrometry for the molecular diagnosis of human breast tumors. Sci Rep 6:201043

6. Dekker TJ, Balluff BD, Jones EA, Schone CD, Schmitt M, Aubele M et al (2014) Multicenter matrix-assisted laser desorption/ionization mass spectrometry imaging (MALDI MSI) identifies proteomic differences in breast-cancer-associated stroma. J Proteome Res 13:4730–4738

7. Pagni F, De Sio G, Garancini M, Scardilli M, Chinello C, Smith AJ et al (2016) Proteomics in thyroid cytopathology: relevance of MALDI-imaging in distinguishing malignant from benign lesions. Proteomics 16:1775–1784

8. Randall EC, Race AM, Cooper HJ, Bunch J (2016) MALDI imaging of liquid extraction surface analysis sampled tissue. Anal Chem 88:8433–8440

9. Buck A, Ly A, Balluff B, Sun N, Gorzolka K, Feuchtinger A et al (2015) High-resolution MALDI-FT-ICR MS imaging for the analysis of metabolites from formalin-fixed, paraffin-embedded clinical tissue samples. J Pathol 237:123–132

10. Casadonte R, Caprioli RM (2011) Proteomic analysis of formalin-fixed paraffin-embedded tissue by MALDI imaging mass spectrometry. Nat Protoc 6:1695–1709

11. Magdeldin S, Yamamoto T (2012) Toward deciphering proteomes of formalin-fixed paraffin-embedded (FFPE) tissues. Proteomics 12:1045–1058

12. Craven RA, Cairns DA, Zougman A, Harnden P, Selby PJ, Banks RE (2013) Proteomic analysis of formalin-fixed paraffin-embedded renal tissue samples by label-free MS: assessment of overall technical variability and the impact of block age. Proteomics Clin Appl 7:273–282

13. Wisniewski JR, Zougman A, Nagaraj N, Mann M (2009) Universal sample preparation method for proteome analysis. Nat Methods 6:359–362

14. Wisniewski JR, Ostasiewicz P, Mann M (2011) High recovery FASP applied to the proteomic analysis of microdissected formalin fixed paraffin embedded cancer tissues retrieves known colon cancer markers. J Proteome Res 10:3040–3049

15. Djidja MC, Claude E, Scriven P, Allen DW, Carolan VA, Clench MR (2017) Antigen retrieval prior to on-tissue digestion of formalin-fixed paraffin-embedded tumour tissue sections yields oxidation of proline residues. Biochim Biophys Acta 1865:901–906

16. Manning JM, Meister A (1966) Conversion of proline to collagen hydroxyproline. Biochemistry 5:1154–1165

17. Kurahashi T, Miyazaki A, Suwan S, Isobe M (2001) Extensive investigations on oxidized amino acid residues in H(2)O(2)-treated cu, Zn-SOd protein with LC-ESI-Q-TOF-MS, MS/MS for the determination of the copper-binding site. J Am Chem Soc 123:9268–9278

18. Xu G, Chance MR (2007) Hydroxyl radical-mediated modification of proteins as probes for structural proteomics. Chem Rev 107:3514–3543

19. Li-En L, Pin-Rui S, Hsin-Yi W, Cheng-Chih H (2018) A simple sonication improves protein signal in matrix-assisted laser desorption ionization imaging. J Am Soc Mass Spectr:1–4

20. Hankin JA, Barkley RM, Murphy RC (2007) Sublimation as a method of matrix application for mass spectrometric imaging. J Am Soc Mass Spectrom 18:1646–1652

21. Yang J, Caprioli RM (2011) Matrix sublimation/recrystallization for imaging proteins by mass spectrometry at high spatial resolution. Anal Chem 83:5728–5734

22. Ji JB, Lu XH, Cai MQ, Xu ZC (2006) Improvement of leaching process of Geniposide with ultrasound. Ultrason Sonochem 13:455–462

23. Djidja MC, Claude E, Snel MF, Francese S, Scriven P, Carolan V et al (2010) Novel molecular tumour classification using MALDI-mass spectrometry imaging of tissue micro-array. Anal Bioanal Chem 39:587–601

24. Kriegsmann M, Longuespee R, Wandernoth P, Mohanu C, Lisenko K, Weichert W et al (2017) Typing of colon and lung adenocarcinoma by high throughput imaging mass spectrometry. Biochim Biophys Acta 1865:858–864

25. Suk KT, Kim DJ (2012) Drug-induced liver injury: present and future. Clin Mol Hepatol 18:249–257
26. Robles-Diaz M, Medina-Caliz I, Stephens C, Andrade RJ, Lucena MI (2016) Biomarkers in DILI: one more step forward. Front Pharmacol 7:1–7
27. Ramm S, Morissey B, Hernandez B, Rooney C, Pennington SR (2015) Mally, a: application of a discovery to targeted LC-MS proteomics approach to identify deregulated proteins associated with idiosyncratic liver toxicity in a rat model of LPS/diclofenac co-administration. Toxicology 331:100–111
28. Gao Y, Cao Z, Yang X, Abdelmegeed MA, Sun J, Chen S et al (2017) Proteomic analysis of acetaminophen-induced hepatotoxicity and identification of heme oxygenase 1 as a potential plasma biomarker of liver injury. Proteomics Clin Appl 11:1–2
29. Castillo MJ, Reynolds KJ, Gomes A, Fenselau C, Yao X (2001) Quantitative protein analysis using enzymatic [^{18}O]water labeling. Anonymous Current Protocols in Protein Science. John Wiley & Sons, Inc., In
30. Committee on Toxicity Testing and Assessment of Environmental Agents, Board on Environmental Studies and Toxicology, Institute for Laboratory Animal Research, Division on Earth and Life Studies, National Research Council (2007) Toxicity testing in the 21st century: A vision and a strategy. In: Anonymous Toxicity Testing in the 21st Century: A Vision and a Strategy, pp 1–196
31. Bruderer R, Bernhardt OM, Gandhi T, Miladinovic SM, Cheng L, Messner S et al (2015) Extending the limits of quantitative proteome profiling with data-independent acquisition and application to acetaminophen-treated three-dimensional liver microtissues. Mol Cell Proteomics 14:1400–1410
32. Gonneaud A, Asselin C, Boudreau F, Boisvert F (2017) Phenotypic Analysis of Organoids by Proteomics. Proteomics 17:. n/a
33. Netzlaff F, Lehr C, Wertz PW, Schaefer UF (2005) The human epidermis models EpiSkin, SkinEthic and EpiDerm: an evaluation of morphology and their suitability for testing phototoxicity, irritancy, corrosivity, and substance transport. Eur J Pharm Biopharm 60:167–178
34. Winget JM, Finley D, Mills KJ, Hugglis T, Bascom C, Isfor RJ et al (2016) Quantitative Proteomic Analysis of Stratum Corneum Dysfunction in Adult Chronic Atopic Dermatitis J Invest Dermatol 136:1732–1735
35. Konstantakou EG, Velentzas AD, Anagnostopoulos AK, Litou ZI, Konstandi OA, Giannopoulou AF et al (2017) Deep-proteome mapping of WM-266-4 human metastatic melanoma cells: From oncogenic addiction to druggable targets. PLoS One 12:e0171512
36. Guran R, Vanickova L, Horak V, Krizkova S, Michalek P, Heger Z et al (2017) MALDI MSI of MeLiM melanoma: Searching for differences in protein profiles. PLoS One 12:e0189305
37. Dowling P, Moran B, McAuley E, Meleady P, Henry M, Clynes M et al (2016) Quantitative label-free mass spectrometry analysis of formalin-fixed, paraffin -embedded tissue representing the invasive cutaneous malignant melanoma proteome. Oncol Lett 12:3296–3304
38. Hoper T, Mussotter F, Haase A, Luch A, Tralau T (2017) Application of proteomics in the elucidation of chemical-mediated allergic contact dermatitis. Toxicol Res 6:595–610
39. Guedes S, Neves B, Vitorino R, Domingues R, Cruz MT, Domingues P (2017) Contact dermatitis: in pursuit of sensitizer's molecular targets through proteomics. Arch Toxicol 91:811–825
40. Ion A, Popa IM, Papagheorghe LML, Lisievici C, Lupu M, Voiculescu V et al (2016) Proteomic approaches to biomarker discovery in cutaneous T-cell lymphoma. Dis Markers 2016:1–8
41. Lundberg KC, Fritz Y, Johnston A, Foster AM, Baliwag J, Gudjonsson JE et al (2015) Proteomics of skin proteins in psoriasis: from discovery and verification in a mouse model to confirmation in humans. Molecular & cellular proteomics: MCP 14:109–119
42. Reindl J, Pesek J, Krüger T, Wendler S, Nemitz S, Muckova P et al (2016) Proteomic biomarkers for psoriasis and psoriasis arthritis. J Proteome 140:55–61

43. Wang J, Suárez-Fariñas M, Estrada Y, Parker ML, Greenlees L, Stephens G et al (2017) Identification of unique proteomic signatures in allergic and non-allergic skin disease. Clin Exp Allergy 47:1456–1467
44. Cretu D, Liang K, Saraon P, Batruch I, Diamandis E, Chandran V (2015) Quantitative tandem mass-spectrometry of skin tissue reveals putative psoriatic arthritis biomarkers. Clin Proteomics 12:1–8
45. Fang J, Wang P, Huang C, Chen M, Wu Y, Pan T (2016) Skin aging caused by intrinsic or extrinsic processes characterized with functional proteomics. Proteomics 16:2718–2731
46. Wang W, Luo J, Sheng W, Xue J, Li M, Ji J et al (2016) Proteomic profiling of radiation-induced skin fibrosis in rats: targeting the ubiquitin-proteasome system. Int J Radiat Oncol Biol Phys 95:751–760
47. Dyer JM, Haines SR, Thomas A, Wang W, Walls RJ, Clerens S et al (2017) Redox proteomic evaluation of oxidative modification and recovery in a 3D reconstituted human skin tissue model exposed to UVB. Int J Cosmet Sci 39:197–205
48. Lee SH, Matsushima K, Miyamoto K, Oe T (2016) UV irradiation-induced methionine oxidation in human skin keratins: mass spectrometry-based non-invasive proteomic analysis. J Proteome 133:54–65
49. Moon E, Park HM, Lee CH, Do S, Park J, Han N et al (2015) Dihydrolipoyl dehydrogenase as a potential UVB target in skin epidermis; using an integrated approach of label-free quantitative proteomics and targeted metabolite analysis. J Proteome 117:70–85
50. Van Eijl S, Zhu Z, Cupitt J, Gierula M, Gotz C, Fritsche E et al (2012) Elucidation of xenobiotic metabolism pathways in human skin and human skin models by proteomic profiling. PLoS ONE:7
51. Chromy BA, Eldridge A, Forsberg JA, Brown TS, Kirkup BC, Elster E et al (2014) Proteomic sample preparation for blast wound characterization. Proteome Sci 12:1–8
52. Kalkhof S, Forster Y, Schmidt J, Schulz MC, Baumann S, Weissflog A et al (2014) Proteomics and metabolomics for in situ monitoring of wound healing. Biomed Res Int 2014:1–12
53. Sabino F, Egli FE, Savickas S, Holstein J, Kaspar D, Rollmann M et al (2018) Comparative Degradomics of porcine and human wound exudates unravels biomarker candidates for assessment of wound healing progression in trauma patients. J Invest Dermatol 138:413–422
54. Taverna D, Pollins AC, Sindona G, Caprioli RM, Nanney LB (2016) Imaging mass spectrometry for accessing molecular changes during burn wound healing. Wound Repair Regen 24:775–785
55. Fadini GP, Menegazzo L, Rigato M, Scattolini V, Poncina N, Bruttocao A et al (2016) NETosis delays diabetic wound healing in mice and humans. Diabetes 65:1061–1071
56. Broszczak DA, Sydes ER, Wallace D, Parker TJ (2017) Molecular aspects of wound healing and the rise of venous leg ulceration: Omics approaches to enhance knowledge and aid diagnostic discovery. Clin Biochem Rev 38:35–55
57. Boxman ILA, Hensbergen PJ, Van Der Schors RC, Bruynzeel DP, Tensen CP, Ponec M (2002) Proteomic analysis of skin irritation reveals the induction of HSP27 by sodium lauryl sulphate in human skin. Br J Dermatol 146:777–785
58. Parkinson, E., Skipp,P., Aleksic, M., Garrow, A., Dadd, T., Hughes, M. et al.: Proteomic analysis of the human skin proteome after In Vivo treatment with sodium dodecyl sulphate. PLoS ONE 9(2014)
59. Hart PJ, Francese S, Woodroofe MN, Clench MR (2013) Matrix assisted laser desorption ionisation ion mobility separation mass spectrometry imaging of ex-vivo human skin. Int J Ion Mobil Spectrom 16:71–83
60. Koppes SA, Engebretsen KA, Agner T, Angelova-Fischer I, Berents T, Brandner J et al (2017) Current knowledge on biomarkers for contact sensitization and allergic contact dermatitis. Contact Dermatitis 77:1–16
61. Sabino F, Hermes O, Egli FE, Kockmann T, Schlage P, Croizat P et al (2015) In vivo assessment of protease dynamics in cutaneous wound healing by degradomics analysis of porcine wound exudates. Mol Cell Proteomics 14:354–370

62. Paulo J (2016) Sample preparation for proteomic analysis using a GeLC-MS/MS strategy. J Biol Methods 3:1–16
63. Test No. 442C: In Chemico Skin Sensitisation -Direct Peptide Reactivity Assay (DPRA), In series:OECD Guidelines for the Testing of Chemicals, Section 4: Health Effects. Published on February 05, 2015 https://dx.doi.org/10.1787/9789264229709-en
64. Wiese S, Reidegeld KA, Meyer HE, Warscheid B (2007) Protein labeling by iTRAQ: a new tool for quantitative mass spectrometry in proteome research. Proteomics 7:340–350
65. Hart PJ, Clench MR (2017) MALDI-MSI of lipids in human skin. Methods Mol Biol 1618:19–36
66. de Macedo CS, Anderson DM, Schey KL (2017) MALDI (matrix assisted laser desorption ionization) imaging mass spectrometry (IMS) of skin: aspects of sample preparation. Talanta 174:325–335
67. Enthaler B, Trusch M, Fischer M et al (2013) MALDI imaging in human skin tissue sections: focus on various matrices and enzymes. Anal Bioanal Chem 405:1159–1170
68. Taverna D, Pollins AC, Sindona G, Caprioli RM, Nanney LB (2015) Imaging mass spectrometry for assessing cutaneous wound healing: analysis of pressure ulcers. J Proteome Res 14:986–996
69. Schmidt A, Bekeschus S, Wende K, Vollmar B, von Woedtke T (2017) A cold plasma jet accelerates wound healing in a murine model of full-thickness skin wounds. Exp Dermatol 26:156–162
70. Lewis EE, Freeman-Parry L, Bojar RA, Clench MR (2018) Examination of the wound healing process using a living skin equivalent (LSE) model and matrix-assisted laser desorption-ionization-mass spectrometry imaging (MALDI-MSI). International Journal of Cosmetic Science (In Press)
71. Ifa DR, Manicke NE, Dill AL, Cooks RG (2008) Latent fingerprint chemical imaging by mass spectrometry. Science 321:805
72. Rowell F, Hudson K, Seviour J (2009) Detection of drugs and their metabolites in dusted latent fingermarks by mass spectrometry. Analyst 134:701–707
73. Wolstenholme R, Bradshaw R, Clench MR, Francese S (2009) Study of latent fingermarks by matrix-assisted laser desorption/ionisation mass spectrometry imaging of endogenous lipids. Rapid Commun Mass Spectrom 23:3031–3039
74. Francese S, Bradshaw R, Denison N (2017) An update on MALDI mass spectrometry based technology for the analysis of fingermarks - stepping into operational deployment. Analyst 142:2518–2546
75. Bandey H, Bowman V, Bleay S, Downham R, Sears VH (2014) In: Bandey H (ed) Fingermark Visualisation Manual. CAST, Home Office, Sandridge, UK
76. Ferguson L, Bradshaw R, Wolstenholme R, Clench MR, Francese S (2011) Two-step matrix application for the enhancement and imaging of latent fingermarks. Anal Chem 83:5585–5591
77. Bailey MJ, Bright NJ, Croxton RS, Francese S, Ferguson LS, Hinder S et al (2012) Chemical characterization of latent fingerprints by matrix-assisted laser desorption ionization, time-of-flight secondary ion mass spectrometry, mega electron volt secondary mass spectrometry, gas chromatography/mass spectrometry, X-ray photoelectron spectroscopy, and attenuated total reflection Fourier transform infrared spectroscopic imaging: an intercomparison. Anal Chem 84:8514–8523
78. Ferguson LS, Creasey S, Wolstenholme R, Clench MR, Francese S (2013) Efficiency of the dry-wet method for the MALDI-MSI analysis of latent fingermarks. J Mass Spectrom 48:677–684
79. Yagnik GB, Kortea AR, Lee YJ (2013) Multiplex mass spectrometry imaging for latent fingerprints. J Mass Spectrom 48:100–104
80. Bailey MJ, Bradshaw R, Francese S, Salter TL, Costa C, Ismail M et al (2015) Rapid detection of cocaine, benzoylecgonine and methylecgonine in fingerprints using surface mass spectrometry. Analyst 140:6254–6259

81. Bradshaw R, Denison N, Francese S (2017) Implementation of MALDI MS profiling and imaging methods for the analysis of real crime scene fingermarks. Analyst 142:1581–1590
82. Kaplan-Sandquist K, LeBeau MA, Miller ML (2015) Evaluation of four fingerprint development methods for touch chemistry using matrix-assisted laser desorption ionization/time-of-flight mass spectrometry. J Forensic Sci 60:610–618
83. Sundar L, Rowell F (2014) Detection of drugs in lifted cyanoacrylate-developed latent fingermarks using two laser desorption/ionisation mass spectrometric methods. Analyst 139:633–642
84. Bradshaw R, Wolstenholme R, Blackledge R, Clench MR, Ferguson L, Francese S (2011) A novel matrix-assisted laser desorption/ionisation mass spectrometry imaging based methodology for the identification of sexual assault suspects. Rapid Commun Mass Spectrom 25:415–422s
85. Bradshaw R, Wolstenholme R, Ferguson LS, Sammon C, Mader K, Claude E et al (2013) Spectroscopic imaging based approach for condom identification in condom contaminated fingermarks. Analyst 138:2546–2557
86. Bradshaw R, Bleay S, Clench MR, Francese S (2017) Direct detection of blood in fingermarks by MALDI MS profiling and imaging. Sci Justice 54:110–117
87. Patel E, Cicatiello P, Deininger L, Clench MR, Marino G, Giardina P et al (2016) A proteomic approach for the rapid, multi-informative and reliable identification of blood. Analyst 141:191–198
88. Deininger L, Patel E, Clench MR, Sears V, Sammon C, Francese S (2016) Proteomics goes forensic: detection and mapping of blood signatures in fingermarks. Proteomics 16:1707–1717
89. Groeneveld G, DePuit M, Bleay S, Bradshaw R, Francese S (2015) Detection and mapping of illicit drugs and their metabolites in fingermarks by MALDI MS and compatibility with forensic techniques. Sci Rep 5:1–13
90. Ferguson LS, Wulfert F, Wolstenholme R, Fonville JM, Clench MR, Carolan VA et al (2012) Direct detection of peptides and small proteins in fingermarks and determination of sex by MALDI mass spectrometry profiling. Analyst 137:4686–4692
91. Flad T, Bogumil R, Tolson J, Schittek B, Garbe C, Deeg M et al (2002) Detection of dermcidin-derived peptides in sweat by ProteinChip technology. J Immunol Methods 270:53–62
92. Rieg S, Seeber S, Steffen H, Humeny A, Kalbacher H, Stevanovic S et al (2006) Generation of multiple stable dermcidin-derived antimicrobial peptides in sweat of different body sites. J Invest Dermatol 126:354–365
93. Baechle D, Flad T, Cansier A, Steffen H, Schittek B et al (2006) Cathepsin D is present in human Eccrine sweat and involved in the Postsecretory processing of the antimicrobial peptide DCD-1L. J Biol Chem 281:5406–5415
94. Wolf R, Voscopoulos C, Winston J, Dharamsi A, Goldsmith P, Gunsior M et al (2009) Highly homologous hS100A15 and hS100A7 proteins are distinctly expressed in normal breast tissue and breast cancer. Cancer Lett 277:101–107
95. Moreira DF, Strauss BE, Vannier E, Belizario JE (2008) Genes up- and down-regulated by dermcidin in breast cancer: a microarray analysis. Genet Mol Res 7:925–932
96. Stewart GD, Skipworth RJE, Pennington CJ, Lowrie AG, Deans DAC, Edwards DR et al (2008) Variation in dermcidin expression in a range of primary human tumours and in hypoxic/oxidatively stressed human cell lines. Br J Cancer 99:126–132
97. Sutton CW, Pemberton KS, Cottrell JS, Corbett JM, Wheeler CH, Dunn MJ et al (1995) Identification of myocardial proteins from two-dimensional gels by peptide mass fingerprinting. Electrophoresis 16:308–316
98. Chen EI, Cociorva D, Norris JL, Yates JR III (2007) Optimization of mass spectrometry-compatible surfactants for shotgun proteomics. J Proteome Res 6:2529–2538
99. Djidja MC, Francese S, Loadman PM, Sutton CW, Scriven P, Claude E et al (2009) Detergent addition to tryptic digests and ion mobility separation prior to MS/MS improves peptide yield and protein identification for in situ proteomic investigation of frozen and formalin-fixed paraffin-embedded adenocarcinoma tissue sections. Proteomics 9:2750–2763

100. Patel E, Clench MR, West A, Marshall S, Marshall N, Francese S (2015) Alternative surfactants for improved efficiency of in situ Tryptic proteolysis of fingermarks. J Am Soc Mass Spectrom 26:862–872

101. Huang HZ, Nichols A, Liu D (2009) Direct identification and quantification of aspartyl succinimide in an IgG2 mAb by RapiGest assisted digestion. Anal Chem 81:1686–1692

102. Yu YQ, Gilar M, Lee PJ, Bouvier ES, Gebler JC (2003) Enzyme-friendly, mass spectrometry-compatible surfactant for in-solution enzymatic digestion of proteins. Anal Chem 75:6023–6028

103. Lee HC, Gaensslen RE (2001) Advances in Fingerprint Technology. CRC Press, Boca Raton, pp 63–104

104. Raiszadeh MM, Ross MM, Russo PS, Schaepper MA, Zhou W, Deng J et al (2012) Proteomic analysis of Eccrine sweat: implications for the discovery of schizophrenia biomarker proteins. J Proteome Res 11:2127–2139

105. Bossers LCAM, Roux C, Bell M, McDonagh AM (2011) Methods for the enhancement of fingermarks in blood. Forensic Sci Int 210:1–11

106. Liumbruno G, D'Alessandro A, Grazzini G, Zolla L (2010) Blood-related proteomics. J Proteome 73:483–507

107. Martin NJ, Bunch J, Cooper HJ (2013) Dried blood spot proteomics: surface extraction of endogenous proteins coupled with automated sample preparation and mass spectrometry analysis J. Am Soc Mass Spectrom 24:1242–1249

108. Beutler E, Waalen J (2006) The definition of anemia : what is the lower limit of normal of the blood hemoglobin concentration ? Blood 107:1747–1750

109. Shen Y, Jacobs JM, Camp DG, Fang R, Moore RJ, Smith RD et al (2004) Ultra-high-efficiency strong cation exchange LC/RPLC/MS/MS for high dynamic range characterization of the human plasma proteome. Anal Chem 76:1134–1144

110. Petibois C, Cazorla G, Cassaigne A, Déléris G (2001) Plasma protein contents determined by Fourier-transform infrared spectrometry. Clin Chem 47:730–738

111. Bollineni RC, Guldvik IJ, Grönberg H, Wiklund F, Mills IG, Thiede B (2015) A differential protein solubility approach for the depletion of highly abundant proteins in plasma using ammonium sulfate. Analyst 140:8109–8117

112. Kamanna SJ, Voelcker HN, Linacre A, Kirkbride KP (2017) Bottom-up in situ proteomic differentiation of human and non-human haemoglobins for forensic purposes by matrix-assisted laser desorption/ionization time-of-flight tandem mass spectrometry. Rapid Commun Mass Spectrom 31:1927–1937

113. Longobardi S, Gravagnuolo A, Funari R, Della Ventura B, Pane F, Galano E et al (2015) A simple MALDI plate functionalization by Vmh2 hydrophobin for serial multi-enzymatic protein digestions. Anal Bioanal Chem 407:487–496

114. De Stefano L, Rea I, De Tommasi E, Rendina I, Rotiroti L, Giocondo M et al (2009) Bioactive modification of silicon surface using self-assembled hydrophobins from Pleurotus ostreatus. Eur Phys J 30:181–185

115. Longobardi S, Gravagnuolo AM, Rea I, De Stefano L, Marino G, Giardina P (2014) Hydrophobin-coated plates as matrix-assisted laser desorption/ionization sample support for peptide/protein analysis. Anal Biochem 449:9–16

116. Premasiri WR, Lee JC, Ziegler LD (2012) Surface-enhanced Raman scattering of whole human blood, blood plasma, and red blood cells: cellular processes and bioanalytical sensing. J Phys Chem 116:9376–9386

117. Boyd S, Bertino MF, Ye D, White LS, Seashols SJ (2013) Highly sensitive detection of blood by surface enhanced Raman scattering. J Forensic Sci 58:753–756

118. Sikirzhytski V, Sikirzhytskaya A, Lednev IK (2011) Multidimensional Raman spectroscopic signatures as a tool for forensic identification of body fluid traces: a review. Appl Spectrosc 65:1223–1232

119. Sikirzhytski V, Sikirzhytskaya A, Lednev IK (2012) Advanced statistical analysis of Raman spectroscopic data for the identification of body fluid traces: semen and blood mixtures. Forensic Sci Int 222:259–265

120. Virkler K, Lednev IK (2008) Raman spectroscopy offers great potential for the nondestructive confirmatory identification of body fluids. Forensic Sci Int 181:1–5
121. Sikirzhytski V, Virkler K, Lednev IK (2010) Discriminant analysis of Raman spectra for body fluid identification for forensic purposes. Sensors 10:2869–2884
122. Virkler K, Lednev IK (2010) Raman spectroscopic signature of blood and its potential application to forensic body fluid identification. Anal Bioanal Chem 396:525–534
123. Li B, Beveridge P, O'Hare WT, Islam M (2014) The application of visible wavelength reflectance hyperspectral imaging for the detection and identification of blood stains. Sci Justice 54:32–38
124. Deininger L, Francese S, Clench MR, Langenburg G, Sears V, Sammon C (2018) Investigation of infinite focus microscopy for the determination of the association of blood with fingermarks. Sci Justice 58(6):397–404
125. Francese S (2019) Criminal profiling through MALDI MS based technologies – breaking barriers towards border free forensic science. Aust J Forensic Sci., https://doi.org/10.1080/00450618.2018.1561949
126. Ramirez T, Daneshian MH, Bois FY, Clench MR et al (2013) T4 report* metabolomics in toxicology and preclinical research. ALTEX 30:209–225

Chapter 5
Sample Treatment for Urine Proteomics

Fernando Sánchez-Juanes and José Manuel González-Buitrago

5.1 Introduction

Urine is a biological fluid of great interest for the search of biomarkers of disease [1]. It can be collected noninvasively in relatively large quantities and is less complex than serum. It has been used mainly to study renal physiology and kidney diseases, although it has also been used for the study of other diseases. Normal urine contains up to 150 mg/24 h protein, with a very variable concentration that depends on the volume of urine, ranging from 2 to 10 mg/dL. These proteins are mainly low molecular weight proteins that have been filtered in the glomerulus and proteins produced in the urinary tract. As urine collects substances from various components of the urinary tract (kidney, bladder) it is a fluid in which biomarkers of urological diseases that affect any of these components can be detected [1]. In addition, by eliminating products from the glomerular filtration of blood plasma, substances

F. Sánchez-Juanes
Instituto de Investigación Biomédica de Salamanca (IBSAL), Complejo Asistencial
Universitario de Salamanca, Salamanca, Spain

Departamento de Bioquímica y Biología Molecular, Universidad de Salamanca, Salamanca,
Spain
e-mail: fsjuanes@usal.es

J. M. González-Buitrago (✉)
Instituto de Investigación Biomédica de Salamanca (IBSAL), Complejo Asistencial
Universitario de Salamanca, Salamanca, Spain

Departamento de Bioquímica y Biología Molecular, Universidad de Salamanca, Salamanca,
Spain

Servicio de Análisis Clínicos/Bioquímica Clínica, Complejo Asistencial Universitario de
Salamanca, Salamanca, Spain
e-mail: buitrago@usal.es

© Springer Nature Switzerland AG 2019
J.-L. Capelo-Martínez (ed.), *Emerging Sample Treatments in Proteomics*, Advances
in Experimental Medicine and Biology 1073, https://doi.org/10.1007/978-3-030-12298-0_5

secreted by damaged organs or tissues can be detected in urine and therefore used as biomarkers of non-urological diseases, such as oncological [2, 3], cardiovascular [4, 5], or rheumatological diseases [6].

Proteomics is the study of the proteins of a biological medium. Comparison of protein patterns of biological fluids between healthy people and patients with a certain disease can be used to discover biological markers of the disease. Urine after serum or plasma is the second biological fluid in which more proteomic studies have been carried out, both to detect the proteins eliminated by it and to look for biomarkers of disease of the urological tract and of systemic diseases. In this chapter, we discuss the aspects related to the collection and treatment of urine for proteomic studies.

5.2   Protein Composition of Urine

Albumin is the main protein in blood plasma and also appears in normal urine. Other proteins present in small amounts are plasma and tubular microglobulins, Tamm-Horsfall protein (uromodulin) produced by epithelial cells of the renal tubule, and proteins of the prostatic, seminal, and vaginal secretions. In healthy people, approximately 30% of the proteins in urine are from plasma, while the other 70% correspond to kidney proteins. Using proteomic methods, a high number of proteins have been detected, more than 1500, including a large proportion of membrane proteins, in the study of Adachi et al. [7] and 3429 in the more recent study of Santucci et al. [8]. Similarly to plasma, the range of protein concentration in urine spans several orders of magnitude.

In general, protein composition of urine is determined by the functionality of the glomerular filtration apparatus, the absorption in the proximal tubule of ultrafiltered proteins, and the ability of the proteolytic machinery of the brush and lysosomal border to degrade filtered proteins. In addition, the generation of certain proteins by the damaged kidney or the lower urinary tract may result in their appearance in urine, either intact or, more likely, as peptide fragments.

Proteinuria is the increase of protein excretion in urine above reference values. The extent of proteinuria and the protein composition of urine depend on the mechanism of kidney damage that leads to protein loss. Proteinuria, which is probable the most frequent and early sign of kidney disease, often precedes other signs or symptoms of kidney disease in months or even years. However, it does not always indicate the existence of a pathological state, so it is a sensitive but not specific marker.

The increased protein excretion in urine can occur by:

- Protein increased in plasma, which causes the capacity of tubular reabsorption to be exceeded
- Increased permeability of the glomerular filtration apparatus
- Decreased reabsorption capacity of the distal convoluted tubule
- Increased endogenous production of proteins by the urinary tract

Proteinuria can be classified according to various criteria; the main ones are the amount excreted and the clinical approach [9]. According to the quantity released, proteinuria can be intense, moderate, or minimal. From the clinical point of view, proteinuria can be symptomatic or asymptomatic, that is, associated or not with symptoms of kidney disease. Likewise, it can be intermittent or persistent. Intermittent proteinuria, which usually results benign, may be transient, functional, or postural. Persistent proteinuria is one that remains over time. According to its origin, it can be prerenal, renal, or postrenal. Depending on the site of renal damage, renal proteinuria is classified as glomerular, tubular, and mixed.

5.3 Requirements for a Sample Preparation Method for Proteomic Studies

The main characteristics that must be requested from a sample preparation method for proteomic analysis are that it should be reproducible, so as to solubilize proteins of all kinds, including hydrophobic ones; that it should avoid aggregation and loss of protein solubility; that it should avoid posttranslational modifications, including enzymatic or chemical degradation of protein sample; that it should eliminate or completely digest nucleic acids and other interfering molecules; and that it should bring the concentrations of proteins of interest to detectable levels.

The methods of preparation of a urine sample depend on the techniques that will be used later for the separation and identification of the proteins. The two main proteomic methodologies for protein separation are solid media such as two-dimensional gel electrophoresis and liquid methods such as HPLC or capillary electrophoresis or combinations thereof in multidimensional methods. The proteomic strategy should also be taken into account, whether it is top-down or bottom-up proteomics.

5.4 Origin of the Urine Sample

Regarding the origin of the samples in the proteomic studies of urine, several possibilities have been employed. A sample of a single individual (male or female), or a sample pool of several individuals, whether men, women, or both sexes, may be used. In their studies on the methods of preparation of urine sample for proteomics, Thongboonkerd et al. used urine specimens from four men and four women, joining them in a pool to examine the difference among various sample preparation protocols [10]. Adachi et al. used pooled urine sample from nine healthy individuals (five male and four female) aged 26–61 years to obtain human urinary proteome [7].

5.5   Considerations About Urine Sample

For proteomic studies, in general, it is not recommended neither the use of protease inhibitors nor the addition of a preservative to prevent bacterial overgrowth. However, this is a controversial issue. Hepburn et al. evaluated the impact of pre-analytical factors on the urine proteome, particularly sample processing time, temperature, and proteolysis [11]. They concluded that control of analytical and pre-analytical variables is crucial for the success of urine analysis, to obtain meaningful and reproducible data. The most critical aspects are the adoption of consistent protocols within the studies. Another important issue in the preparation of urine samples for proteomic studies is the removal of cells and debris to avoid contamination with proteins from red blood cells, white blood cells, or epithelial cells. This is done by centrifuging urine at low revolutions.

As noted above, sample preparation before its analysis is one of the critical steps in proteomic studies. This preparation should be as simple as possible to increase reproducibility. Likewise, it should be ensured that during sample preparation proteins undergo the least possible modifications.

Normal human urine has a much diluted protein concentration with a high-salt content, which interferes with proteomic analysis. Thus, an initial step in the handling of urine samples should be to concentrate and to eliminate salts. To this end, most commonly used methods are precipitation, lyophilization, ultracentrifugation, and centrifugal filtration. With these methods salts were eliminated and all the proteins present in the urine were concentrated.

As pointed out by Thongboonkerd et al., the results obtained with these methods are usually complementary, so it is necessary to combine several of them to obtain the greatest amount of qualitative and quantitative information [10]. More recently, Paulo et al. compared four different extraction methods to determine the most effective method: ethanol precipitation, lyophilization, microconcentrators, and C4 trapping column [12]. They found that out of the top 100 proteins identified, 89% were found in all four methods. As a conclusion, the choice of the sample preparation method should be based on the objectives to be achieved, the techniques of separation and identification of proteins to be used, and the cost of the analysis to be done.

5.5.1   Precipitation

Precipitation with different organic compounds has been used for a long time to isolate/concentrate urinary proteins. Acetic acid, acetone, acetonitrile, ammonium sulfate, chloroform, ethanol, and methanol with various final concentrations are added to the urine. The precipitant is isolated by centrifugation and resuspended with a solubilizing buffer. Depending on the solvent or combination of solvents that will be used, different groups of proteins could be precipitated. Thongboonkerd et al. used various combinations of protein precipitating agents in a urine study [10].

They concluded that there is no single perfect protocol that can be used to examine the entire urinary proteome as each method has both advantages and disadvantages compared to the others: all of them are indeed complementary. Khan and Packer developed a urine preparation method for two-dimensional gel electrophoresis which involves precipitation of proteins with simultaneous desalting [13]. Acetonitrile precipitation produced 2D gel separations with the highest resolution and greatest number of protein spots compared to precipitation by other organic solvents.

5.5.2   Lyophilization

The urine is lyophilized until dry and then resuspended in the solubilizing buffer. The resuspension is dialyzed against deionized water, lyophilized again, and then resuspended with the solubilizing buffer. Lyophilization has been shown in some studies [10, 14] to provide the best quantitative yield, since it is the procedure with the least protein losses. However, it has the disadvantage that, while concentrating proteins, it also concentrates salts.

5.5.3   Ultracentrifugation

Urinary proteins can be isolated by ultracentrifugation at 200,000 × g at 4 °C for 2 h. The supernatant is discarded and the pellet is resuspended with solubilizing buffer. The resuspension is dialyzed against deionized water, lyophilized, and then resuspended with the solubilizing buffer.

5.5.4   Centrifugal Filtration

Urine is spun at 12000 × g using a 10 kDa cutoff centrifugal column at 4 °C until approximately 1/30 of initial volume remained. Concentrated urine is dialyzed against deionized water, lyophilized, and then resuspended with the solubilizing buffer. Lafitte et al. for the first time presented an optimized preparation of urine samples for two-dimensional gel electrophoresis that uses centrifugal filtration [15]. They used 24 h or morning urine samples collected in the presence of preservative thymol. Samples were centrifuged at 3000 g for 10 min in swing buckets to discard cellular debris. The supernatant was precipitated with protamine sulfate to get rid of nucleic acids. The sample was centrifuged at 10,000 g for 20 min. Free glycosaminoglycans were precipitated with 0.1 N HCl until pH reached 3.0. The solution was then dialyzed for 48 h against water and subsequently filtrated on 0.22 μm filters. The solution was then concentrated at 20 °C by centrifugal filtration. The procedure was very cumbersome, and no depletion or enrichment strategies were used. As pointed out, centrifugal filtration showed some protein losses [14].

5.5.5 Other Methods

Jesus et al. have developed a new ultrafast ultrasonic-based method for shotgun proteomics as well as label-free protein quantification in urine samples [16]. The method first separates the urine proteins using nitrocellulose-based membranes and the proteins are in-membrane digested using trypsin. Overall, the sample treatment pipeline comprising separation, digestion, and identification is done in just 3 h.

5.6 Timing of Sample Collection

Another issue that needs to be addressed in urinary proteomic studies is the timing of sample collection. Urine specimens used in clinical laboratories are basically of three types: first morning, random, and timed. There has been extensive discussion about the most appropriate urine sample to use for the protein excretion research. In most urinary proteomic studies, the urine samples used were first morning and random. However, in the Standard Protocol for Urine Collection and Storage, applicable for the analysis of soluble urine proteins, not for exosome analysis, published in the web of the Human Kidney and Urine Proteome Project (http://www. hkupp.org), the second morning urine is the most recommendable sample. Twenty-four-hour urine is not recommended because of the risk of contamination during collection.

5.7 Removal of the Most Abundant Proteins and Enrichment of the Less Abundant Ones

As noted above, range of protein concentrations in urine spans several orders of magnitude in biological fluids. Blood plasma proteins, such as albumin, transferrin, α_1-antitrypsin, and immunoglobulins, are the most abundant proteins of the normal proteome of urine. Due to the masking produced by these most abundant proteins over the less abundant ones, effective proteomic analyses require either removal of abundant proteins or enrichment of the less abundant ones. Depletion of highly abundant proteins is the most often used approach for identification of low-concentration proteins. Depletion of high-abundant plasma proteins from urine is one approach that would simplify sample complexity and improve the chances of finding potential biomarkers. In many cases, selective removal of the most abundant proteins allows better detection of less abundant ones.

The main methods to remove the most abundant proteins are affinity methods (dyes, AG proteins, peptide ligands), size methods (centrifugal ultrafiltration), and methods with antibodies (monoclonal or polyclonal).

Immunodepletion has been the preferred method to eliminate highly abundant proteins. However, this method has the risk of eliminating proteins that are bound to the abundant proteins and thus will be lost. It is well known that highly abundant proteins, like immunoglobulins, could be associated or could bind to several other proteins, which will be removed together with the immunoglobulins.

In 2009, Kushnir et al. published a work in which they performed for the first time a depletion strategy for removing high-abundant proteins from human urine [17]. They used a commercial plasma protein depletion kit. The multiple affinity removal sample (MARS) column for the depletion of six high-abundant proteins contains affinity binders for the depletion of albumin, transferrin, haptoglobin, IgG, IgA, and α-1 antitrypsin. The results showed that the number of low-abundant proteins identified in urine after depletion increased nearly 2.5-fold. A year later, Magagnotti et al. carried out a comparison of three depletion approaches: two commercial immune-affinity chromatographic columns and a homemade column to remove some of the highly abundant proteins in urine [18]. As authors noted, these approaches had been designed for serum/plasma analysis but were then used in urine because a remarkable amount of blood-derived proteins is also found in this biological fluid. The results obtained have led the authors to conclude that the sequential procedure of urine samples using multiprotein immune-affinity depletion represents a valid tool for simplifying two-dimensional electrophoresis analysis of the urine proteome.

More recently, Filip et al. compared different commercially available depletion strategies for the enrichment of low-abundant proteins in human urine [19]. They used three immunodepletion commercial kits and an ion-exchange commercial kit. The results obtained showed satisfactory reproducibility in protein identification as well as protein abundance. Comparison of the depletion efficiency between the unfractionated and fractionated samples and the different depletion strategies showed efficient depletion in all cases, except for the ion-exchange kit. The depletion efficiency was found slightly higher in normal than in chronic kidney disease samples and normal samples yielded more protein identifications than chronic kidney disease samples. Nevertheless, these depletion strategies did not yield a higher number of identifications in neither the urine from normal nor chronic disease patients. They conclude that when analyzing urine in the context of chronic kidney disease biomarker identification, no added value of depletion strategies can be observed, and analysis of unfractionated starting urine appears to be preferable.

ProteoMiner is a technique that allows both the simultaneous depletion of highly abundant proteins and enrichment of low-abundant ones [20]. It uses a library of hexapeptides that are created by combinatorial synthesis and are bound to a chromatographic support (polyacrylamide beads). The result is a large library of hexapeptide ligands that act as unique ligands of the proteins. The population of beads has such diversity that there is a binding partner for most, if not all, of the proteins in a sample. Each support has an equivalent binding capacity; that is, it is equal to the capacity of union of the very abundant and little abundant proteins. Because there are a limited number of protein binding sites, very abundant proteins soon reach saturation, while low-abundant proteins continue to bind. Very abundant proteins that reach saturation stop binding and the excess of unbound protein molecules is removed by washing. After the elution, the medium- and low-abundant proteins are enriched and the very abundant ones are reduced. This allows the detection and identification of low-abundant proteins.

Castagna et al. used ProteoMiner to analyze human urine proteome [21]. They noted that this technique allows exploring the hidden, or low-abundant, proteome, that is, the very large part of the proteome that has escaped detection up to the time of the study. This approach provides detection of more than 400 different gene products. Later, Candiano et al. also used combinatorial ligand libraries for urine proteome analysis [22]. They presented a simplification of the method, with a single elution of captured proteins to reduce the workload. This single step elutes almost quantitatively the adsorbed proteins, thus ensuring a full recovery. In 2012 Candiano et al. presented a novel elution system for the human urinary proteome after being captured with combinatorial peptide ligand libraries [23]. When eluting with this system, at least 3300 spots were visualized in a bidimensional map. More recently, the same group [8] presented an analysis of the human urine proteome in which a combination of analytical procedures with the use of ProteoMiner and identification of proteins by an Orbitrap Velos MS rendered a characterization of 3429 proteins; a relevant part (1724) being detected for the first time in urine.

5.8 Fractionation of the Sample and Sample Preparation for Specific Proteins

The main objective of proteomic analysis is the detection and identification of the largest possible number of proteins present in a given medium. In most biological media, the concentration range of proteins present is very large; this is the case of urine.

One option in urine preparation is sample fractionation. The reason for the previous fractionation of the samples is the great difference of properties of proteins, like mass, isoelectric point, degree of hydrophobicity, and posttranslational modifications. Thongboonkerd et al. presented a method for enrichment of basic/cationic proteins in urine using ion-exchange chromatography and batch adsorption [24].

Phosphorylation is a common posttranslational modification of proteins that involves the reversible attachment of phosphate groups to the side chain of specific amino acids. The most studied compounds are phosphomonoesters of serine, threonine, and tyrosine that are stable in acid solutions, and thus sample preparation in acidic media is most often used in proteomics. The isolation/enrichment of phosphorylated peptides from protein digests is a crucial step for the sequencing and identification of peptides from phosphorylation sites by MS.

For the enrichment of glycoproteins and glycopeptides, lectins are used, which are proteins that bind carbohydrates with specific structures. The main lectins used are concanavalin A and wheat germ agglutinin. Huo et al. developed a sequential separation method for isolation of N-glycoproteins that isolates urinary soluble proteins and extracellular vesicle proteins via stepwise ultrafiltration based on their obvious size difference [25].

5.9   Urinary Exosomes

Exosomes have been used for biomarkers searching in biological fluids, including urine [26]. Exosomes are small vesicles formed as part of the endosomal pathway that contains cellular material surrounded by a lipid bilayer derived from the plasma membrane. Exosomes are released by most cells of the human body and can be isolated from all biological fluids including urine. Urinary exosomes have been suggested as alternative materials to identify useful biomarkers. Exosomes secreted from epithelial cells lining the urinary tract might reflect the cellular processes associated with the pathogenesis of their donor cells.

Ultracentrifugation is the most used technique for the isolation of exosomes [27–29]. Other techniques also used are filtration, precipitation, affinity purification, and use of microfluidics. It is important to note that these techniques provide different results in terms of recovery and size of exosomes and their corresponding proteins. For most procedures low- and moderate-speed centrifugation of urine is used previously to remove cells or debris. A typical basic procedure for urine exosome isolation is ultracentrifugation at 200,000 x g for 1 h yielding an exosomal pellet that can be resuspended for later use.

5.10   Conclusions

Sample treatment of biological fluids for its proteomics analysis is of great importance for obtaining results that are useful in the search for biological markers of disease. Methods of preparation of urine sample depend on the techniques that later will be used for separation and identification of proteins. For proteomic studies, in general, it is not recommended neither the use of protease inhibitors nor the addition of a preservative. Removal of cells and debris to avoid contamination is a required step, as well as to concentrate proteins and eliminate salts. Finally, removal of the most abundant proteins and enrichment of the less abundant proteins are the steps that can be taken as an option to detect as many proteins as possible.

References

1. González-Buitrago JM, Ferreira L, Lorenzo I (2007) Urinary proteomics. Clin Chim Acta 375:49–56
2. Wang W, Wang S, Zhang M (2017) Identification of urine biomarkers associated with lung adenocarcinoma. Oncotarget 8:38517–38529
3. Jedinak A, Loughlin KR, Moses MA (2018) Approaches to the discovery of non-invasive urinary biomarkers of prostate cancer. Oncotarget 9:32534–32550
4. Htun NH, Magliano DJ, Zhang ZT et al (2017) Prediction of acute coronary syndromes by urinary proteome analysis. PLoS One 12:e0172036
5. Rothlisberger S, Pedroza-Diaz J (2017) Urine protein biomarkers for detection of cardiovascular disease and their use for the clinics. Expert Rev Proteomics 14:1091–1103
6. Kang MJ, Park YJ, Yon S, Yoo SA, Choi S, Kim DH, Cho CS, Yi EC, Hwang D, Kim WU (2014) Urinary proteome profile predictive of disease activity in rheumatoid arthritis. J Proteome Res 13:5206–5217
7. Adachi J, Kumar C, Zhang Y, Olsen JV, Mann M (2006) The human urinary proteome contains more than 1500 proteins, including a large proportion of membrane proteins. Genome Biol 7:R80
8. Santucci L, Candiano G, Petretto A, Bruschi M, Lavarello C, Inglese E, Righetti PG, Ghiggeri GM (2015) From hundreds to thousands: widening the normal human Urinome (1). J Proteome 115:53–62
9. Julian BA, Suzuki M, Suzuki Y, Tomino Y, Spasovski G, Novak J (2009) Sources of urinary proteins and their analysis by urinary proteomics for the detection of biomarkers of disease. Proteomics Clin App 3:1029–1043
10. Thongboonkerd V, Chutipongtanate S, Kanlaya R (2006) Systematic evaluation of sample preparation methods for gel-based human urinary proteomics: quantity, quality and variability. J Proteome Res 5:183–191
11. Hepburn S, Cairns DA, Jackson D (2015) An analysis of the impact of pre-analytical factors on the urine proteome: sample processing time, temperature, and proteolysis. Proteomics Clin App 9:507–521
12. Paulo JA, Vaezzadeh AR, Conwell DL et al (2011) Sample handling of body fluids for proteomics. In: Ivavov A, Lazarev A (eds) Sample preparation in biological mass spectrometry. Springer, Dordrecht, pp 327–360
13. Khan A, Packer NH (2006) Simple urinary sample preparation for proteomic analysis. J Proteome Res 5:2824–2838
14. Lee RS, Monigatti F, Briscoe AC, Waldou Z, Freeman MR, Steen H (2008) Optimizing sample handling for urinary proteomics. J Proteome Res 7:4022–4030
15. Laffite D, Dussol B, Andersen S, Vazi A, Duouy P, Jensen ON, Berland Y, Verdier JM (2002) Optimized preparation of urine samples for two-dimensional electrophoresis and initial application to patient samples. Clin Biochem 25:581–589
16. Jesus JR, Santos HM, López-Fernández H, Lodeiro C, Zezzi Arruda MA, Capelo JL (2018) Ultrasonic-based membrane aided sample preparation of urine proteomes. Talanta 178:864–869
17. Kushnir MP, Rockwood AL, Crockett DK (2009) A depletion strategy for improved detection of human proteins from urine. J Biomol Tech 20:101–108
18. Magagnotti C, Felmo I, Carletti RM, Ferrari M, Bachi A (2010) Comparison of different depletion strategies for improving resolution of the human urine proteome. Clin Chem Lab Med 48:531–535
19. Filip S, Vougas K, Zoidakis J, Latosinska A, Mullen W, Spasovski G, Mischak H, Vlahou A, Jankowsky J (2015) Comparison of depletion strategies for the enrichment of low-abundance proteins in urine. PLoS One 10:0133773
20. Righetti PG, Castagna A, Antonioli P, Boschetti E (2005) Prefractionation techniques in proteome analysis: the mining tool of the third millennium. Electrophoresis 26:297–319

21. Castagna A, Cecconi D, Sennels L et al (2005) Exploring the hidden human urinary proteome via ligand library beads. J Proteome Res 4:1919–1930
22. Candiano G, Dimuccio V, Bruschi M et al (2009) Combinatorial peptide ligand libraries for urine proteome analysis: investigation of different elution systems. Electrophoresis 30:2405–2411
23. Candiano G, Santucci L, Bruschi M, Petretto A, D'Ambrosio C, Scaloni A, Righetti PG, Ghiggeri GM (2012) "Cheek-to-cheek" urinary proteome profiling via combinatorial peptide ligand libraries: a novel, unexpected elution system. J Proteome 75:796–805
24. Thongboonkerd V, Semangoen T, Chutipongtanate S (2007) Enrichment of the basic/cationic proteome using ion exchange chromatography and batch adsorption. J Proteome Res 6:1209–1214
25. Huo B, Chen M, Chen J, Li Y, Zhang W, Wang J, Qin W, Qian X (2018) A sequential separation strategy for facile isolation and comprehensive analysis of human urine N-glycoproteome. Anal Bioanal Chem 410:7305. https://doi.org/10.1007/s00216-018-1338-6
26. Pisitkun T, Shen RF, Knepper MA (2004) Identification and proteomic profiling of hexosomes in human urine. Proc Natl Acad Sci U S A 101:13368–13373
27. Street JM, Koritzinsky EH, Glispie DM, Yuen PST (2017) Urine exosome isolation and characterization. Methods Mol Biol 1641:413–423
28. Street JM, Koritzinsky EH, Glispie DM, Star RA, Yuen PST (2017) Urine exosomes: an emerging trove of biomarkers. Advan Clin Chem 78:103–122
29. Pitto M, Corbetta S, Raimondo F (2015) Preparation of urinary exosomes: methodological issues for clinical proteomics. Methods Mol Biol 1243:43–53

Chapter 6
A Method for Comprehensive Proteomic Analysis of Human Faecal Samples to Investigate Gut Dysbiosis in Patients with Cystic Fibrosis

Griet Debyser, Maarten Aerts, Pieter Van Hecke, Bart Mesuere,
Gwen Duytschaever, Peter Dawyndt, Kris De Boeck, Peter Vandamme,
and Bart Devreese

6.1   Introduction

The human gastrointestinal tract hosts a diverse microbial ecosystem that provides functionalities such as regulation of host energy metabolism, pathogen resistance and inflammatory immune responses. The gut microbiota produce essential end products, such as vitamins and short-chain fatty acids [1–4] including butyrate which serves as the preferred energy source for colonocytes and has anti-inflammatory effects [5–9]. In several colon diseases, previous studies demonstrated a decrease in the abundance and diversity of intestinal bacteria of the phylum *Firmicutes* and more specifically the butyrate-producing bacteria of *Clostridium* clusters IV and XIVa [10–14].

Patients with cystic fibrosis (CF) mainly suffer from persistent lung infections and are repeatedly treated with antibiotics, thereby disturbing the beneficial host-microbiota relationship [15]. In addition, CF also affects the function of pancreas and intestine as the thick mucus in the small intestine decreases the absorption of nutrients and bile acids. Moreover, fat-rich diets combined with a decreased release of digestive enzymes by the pancreatic duct result in a different alimentary environment in which the microbial system resides. Furthermore, intestinal obstruction due to abnormal mucus secretions causes abdominal discomfort [16]. All these anomalies

G. Debyser · M. Aerts · P. Van Hecke · G. Duytschaever · P. Vandamme · B. Devreese (✉)
Ghent University, Department of Biochemistry and Microbiology, Ghent, Belgium
e-mail: bart.devreese@ugent.be

B. Mesuere · P. Dawyndt
Ghent University, Department of Applied Mathematics, Computer Science and Statistics, Ghent, Belgium

K. De Boeck
University Hospital of Leuven, Department of Pediatrics, Leuven, Belgium

© Springer Nature Switzerland AG 2019
J.-L. Capelo-Martínez (ed.), *Emerging Sample Treatments in Proteomics*, Advances in Experimental Medicine and Biology 1073, https://doi.org/10.1007/978-3-030-12298-0_6

have major consequences for the composition of the gut microbiota, cause intestinal inflammation and affect the general well-being of patients with CF [17–21]. Consequently, patients with cystic fibrosis have an altered structural and functional gut microbiota. In order to address the actual effect of CF pathogenesis and concomitant treatment on gut microbiota and general gut health, the use of omics technologies is required.

The aim of this study was to evaluate a shotgun metaproteomic pipeline for the analysis of the protein fraction of faecal samples retrieved from patients with cystic fibrosis in comparison with their unaffected siblings. Many studies investigating the gastrointestinal (GI) microbiota have been conducted on enriched microbial fractions of faecal samples only revealing a snapshot of the GI microbiota and the host-microbiota interactions [22, 23]. Unfortunately, the use of an extraction method that is based on differential centrifugation to enrich the microbial cells could lead to a poor representation of the actual microbiota in the gastrointestinal tract. In addition, by ignoring the host proteome, information about important proteins reflecting host health status are lost. On the other hand, one would expect that bacterial enrichment gives a deeper insight in the bacterial proteome and removes unwanted non-protein materials and small molecules that interfere with the mass spectrometric analysis. We here used a nano-LC-MS/MS method to evaluate two extraction methods commonly used in metaproteomic studies and discuss the results in view of their applicability in a large-scale study on a larger cohort of samples.

6.2 Materials and Methods

6.2.1 Chemicals

Chemicals and facilities for SDS-PAGE were obtained from Bio-Rad (Nazareth Eke, Belgium). Dithiothreitol (DTT) and iodoacetamide (IAA) were purchased from Fluka (Buchs, Switzerland). HPLC-grade acetonitrile (ACN) was obtained from Biosolve (Valkenswaard, the Netherlands), formic acid from Panreac (Barcelona, Spain) and sequencing grade modified trypsin from Promega (Madison, WI, USA). Caffeine, L-methionyl-arginyl-phenylalanyl-alanine acetate·H_2O (MRFA) and UltraMark for calibrating the FT-ICR and LTQ mass analysers were from Thermo Fisher Scientific (Waltham, MA, USA). All other solvents for mass spectrometric sample preparation and HPLC were from Biosolve (Valkenswaard, The Netherlands).

6.2.2 Study Material

Faecal material for evaluating the metaproteomic method was obtained from a 17-year-old healthy female. The faecal samples for the replicates were obtained from a 2-year-old female toddler with CF (sample 23.1) and her 5-year-old sister (sample 24.1). The parents received written information containing the details of the study and gave verbal permission. Hereafter, the Ethics Committee of the University of Leuven, Belgium, approved this verbal informed consent (no. ML4698).

6.2.3   Sample Preparation for the Faecal Proteome Fraction (FPF) and the Enriched Bacterial Fraction (EBF)

Faecal samples were stored frozen at −80 °C and were exposed to multiple freeze-thaw cycles prior to analysis. Microbial cells were extracted from the faecal material by an adapted protocol from Verberkmoes et al. [22]. Unprocessed faecal material was thawed at room temperature, and 1 gram (wet weight) was homogenised in 9 ml of extraction buffer, which consists of 50 mM Tris-HCl/10 mM $CaCl_2$ (pH 7,6) and protease inhibitor cocktail (Complete, Roche Diagnostics Corporation, Indianapolis, IN, USA) (1 tablet per 25 ml). Twenty ml of extraction buffer was added, followed by a low-speed centrifugation (15 min, 4 °C at 400 × g) in a bench-top centrifuge SL16R (Thermo Fisher Scientific, Waltham, USA) for the removal of large, nonbacterial particles. The bacterial fraction (EBF) is subsequently collected by high-speed centrifugation (15 min, 4 °C at 4.000 × g). The supernatant is used for the analysis of the faecal proteome fraction (FPF). The EBF pellet was dissolved in 15 ml 50 mM Tris HCl/10 mM $CaCl_2$ pH 7,6 and 600 µl protease inhibitor cocktail (25×). The bacteria in the EBF and FPF were lysed by sonication on ice for 3 times 30 seconds using a Branson Digital Sonifier Model 250-D operating at an amplitude of 30% and a 1 second on, 1 second off cycle. The sample was subsequently clarified by centrifugation for 15 min at 4000 × g in a benchtop centrifuge SL16R (Thermo Fisher Scientific, Waltham, USA) to remove cell debris. The two protein fractions (EBF and FPF) were transferred to Vivaspin centrifugal devices with a 10-kDa cut-off (Vivascience, Binbrook, UK) and centrifuged at 3000 × g to reduce the volumes to one ml.

6.2.4   Sample Preparation for the Whole Protein Fraction (WPF)

The protein content was extracted from the faecal material by an adapted protocol from Kolmeder et al. [24]. Briefly, the same sample preparation is used as for the whole protein fraction (WPF), but after the low-speed centrifugation, the entire supernatant was sonicated to lyse the bacterial cells in the presence of the host fae-cal proteome. Subsequently, the whole proteome sample was clarified by high-speed centrifugation (15 min, 4 °C at 4000 × g) to remove cell debris. The protein solutions were filtered and concentrated to 1 ml with a 10-kDa cut-off Vivaspin centrifugal device (Vivascience, Binbrook, UK).

6.2.5 Sample Preparation for the Proof-of-Concept Study on One Patient and Its Unaffected Sibling

Unprocessed faecal material from the patient and her unaffected sibling was thawed at room temperature, and 0.5 g (wet weight) was homogenised in 9 ml of extraction buffer, which consists of 50 mM Tris-HCl/10 mM $CaCl_2$ (pH 7.8) and protease inhibitor cocktail (Roche Diagnostics Corporation, Indianapolis, IN, USA) (1 tablet per 25 ml). This was followed by a low-speed centrifugation (30 min, 4 °C at 400 × g) for the removal of large, nonbacterial particles. The supernatants were sonicated to lyse the bacterial cells and subsequently clarified by high-speed centrifugation (60 min, 4 °C at 6000 × g) to remove cell debris. The protein solutions were filtered using a 10-kDa cut-off membrane filter (Vivascience, Binbrook, UK) by centrifugation at 3000 × g to reduce the volumes to 1 ml.

6.2.6 1D SDS-PAGE

First, 20 μl of the protein extract was mixed with 20 μl Laemmli buffer (1 ml Laemmli +87 μl β-mercaptoethanol), and after boiling for 10 min, 20 μl of the reduced sample was loaded on a 12.5% Tris-Glycine gel for 1D SDS-PAGE. A molecular weight (MW) marker (Precision Plus Protein Standards, Bio-Rad, Nazareth Eke, Belgium) was also loaded for band identification. The loaded gel was electromobilised at 125 V; the gel was then fixed and stained overnight with Coomassie Brilliant Blue G-250. After destaining in 30% methanol to obtain a clear background, the complete gel lane was cut out in twelve bands for the evaluation of the extraction protocol or five bands for the replicate study. The different bands were treated for in-gel trypsin digestion. Therefore, gel bands were first destained by two consecutive washes in 150 μl buffer containing 200 mM ammonium bicarbonate in 50% (v/v) acetonitrile in water for 30 min at 30 °C. After drying the gel pieces in a SpeedVac (Thermo Savant, Holbrook, USA), 10 μl trypsin solution (0.002 μg/μl in a 50 mM NH_4HCO_3 buffer solution) was added and allowed to be absorbed by the gel for 40 min on ice. Gel bands were completely immersed by adding additional buffer solution and incubated at 37 °C. After overnight digestion, tryptic peptides were extracted by two consecutive washes in 50 μl buffer (60% (v/v) acetonitrile, 0.1% (v/v) formic acid in water). Pooled peptide extracts were subsequently dried in a SpeedVac, dissolved in 15 μl 2% (v/v) acetonitrile, 0.1% (v/v) formic acid in water and stored at −20 °C until further analysis.

6.2.7 LC-MS Analysis

Tryptic peptides from the different gel slices were individually analysed via a fully automated LC-MS/MS setup. Peptides were separated on an Agilent 1200 chromatographic system (Agilent, Santa Clara, CA, USA.). The eluting peptides were online measured on a LTQ-FT Ultra mass spectrometer, which is a fully integrated hybrid mass spectrometer consisting of a linear ion trap (LTQ) mass spectrometer combined with a Fourier transform ion cyclotron resonance (FT-ICR) mass spectrometer (Thermo Fisher Scientific, Waltham, MA, USA). Two μL of a peptide solution was initially concentrated and desalted on a ZORBAX 300SB-C18 trapping column (5 mm × 0.3 mm, Agilent) at a 4 μl/min flow rate using 2% (v/v) acetonitrile, 0.1% (v/v) formic acid in water. After valve switching, the peptides were injected on a ZORBAX 300SB-C18 analytical column, 150 mm × 75 μm (Agilent, Santa Clara, CA, USA), and separated by a 35-minute linear gradient ranging from 2% (v/v) to 80% (v/v) acetonitrile, 0.1% (v/v) formic acid in water at a 300 nL/min flow rate. The LC effluent was introduced online to a TriVersa NanoMate ESI source (Advion, Ithaca, NY, USA), working in nano-LC mode and equipped with D-chips whereon a 1.7 kV voltage was supplied. The LTQ-FT Ultra mass spectrometer was tuned and calibrated with caffeine, MRFA and UltraMark before measurement. The FT-ICR mass analyser acquired MS scans at a resolution of 100,000 during the LC separation. The three most intense precursor peptides for each MS scan were automatically selected and fragmented by the LTQ ion trap mass analyser, and after two occurrences, precursor masses were excluded for 90 seconds.

6.2.8   Peptide Identification

The raw LC-ESI-FT-MS/MS data were processed using Mascot Daemon v2.3.2 and spectra were searched against the NCBI database (downloaded 15 September 2013, 32,321,419 sequences).[1] Carbamidomethyl (C) was selected as a fixed modification, oxidation (M) as a variable modification, and one missed cleavage was allowed with trypsin as the cleaving agent. Decoy database searches were done simultaneously with a tolerance of 10 ppm for the precursor ion and 0.3 Da in the MS/MS mode. The Mascot DAT files were loaded in Scaffold (Proteome Software Inc., Portland, OR, version 4.2.1), and the MS/MS data were additionally identified using X! Tandem (The GPM, thegpm.org; version CYCLONE 2010.12.01.1). X! Tandem was set up to search a subset of the reversed concatenated database, containing only the proteins identified by Mascot, using the same parameters as Mascot, but glutamate and glutamine to PyroGlu, ammonia loss of the N-terminus, oxidation of methionine and carbamidomethylation of cysteine were specified as variable modifications. Scaffold was used to validate MS/MS-based peptide and protein identifications. Peptide identifications were accepted if they could be established at greater than 95% probability. Peptide probabilities from Mascot and X! Tandem were

---

[1] Since 2013, the database has been dramatically increased, e.g. by inclusion of several metagenomics projects. Therefore, the number of identified peptides/proteins could probably be improved by researching data against recent updates of the database. However, the aim of this paper is to compare protocols, and the main conclusions do not depend on the database used.

assigned respectively by the Scaffold Local FDR algorithm and Peptide Prophet algorithm [25]. Protein identifications were accepted if they could be established at greater than 99% probability and contained at least two identified peptides. Protein probabilities were assigned by the Protein Prophet algorithm [26]. Proteins that contained similar peptides and could not be differentiated based on MS/MS analysis alone were grouped to satisfy the principles of parsimony. Proteins sharing significant peptide evidence were grouped into protein clusters. The data from different samples were normalised in Scaffold based on the sum of the "Unweighted Spectrum Count" for each MS sample and scaled so that they were all the same. The scaling factor for each sample was subsequently applied to each protein group.

6.2.9   Biodiversity and Functional Analysis

In the Mascot DAT files, the false discovery rate (FDR) was adjusted to less than 1%, and the ion score cut-off was set to a minimum of 30. The sequence with the highest score per spectrum passing the filtering criteria was analysed using Unipept (http://unipept.ugent.be), providing interactive treemaps or sunburst plots that give insight in the biodiversity of the samples [27]. The advanced search options were used as well as the advanced missed cleavage handling and the equation of I = L. A minimum of four distinct peptides was set as a filter, to determine if a taxonomy level is present. The normalised spectral counts from the three technical replicates from the patient and the unaffected sibling were statistically analysed with Student's t-test.

Sequence data of identified proteins with minimum two different peptides in Scaffold were extracted from the NCBI website and uploaded as a multi-FASTA file to the Blast2GO software. Gene ontology (GO) annotation and enrichment analysis were done using Blast2GO [28]. Briefly, the proteins were mapped against the functional information (GO terms) using the standard parameters. Next, an annotation step was performed using E-value Hit Filter $1.0E^{-6}$, annotation cut-off value of 55 and a GO weight of five. The enzyme code (EC) and InterProScan search were performed [29]; the ANNEX function was used to enhance the GO annotation [30]. For visualisation of the functional information (GO categories: molecular function and biological process), the analysis tool of the Blast2GO software was used. Finally, KEGG pathway maps and corresponding information were downloaded from the KEGG Pathway Database through Blast2GO (http://www.genome.jp/kegg/pathway.html) [31]. In the replicate study, the GO terms were also searched with Blast2GO. The list of proteins from the patient was tested in comparison with the unaffected sibling for enrichment of GO terms using a two-tailed Fisher's exact test method with a filter value of 0.05 in the FDR term filter mode.

6.3   Results

6.3.1   Evaluation of the Extraction Methods

Proteins were extracted from 1 gram of faecal material from a 17-year-old healthy person using two different extraction methods. The first extraction method provides the enriched bacterial fraction and the faecal proteome fraction. The second method provides the whole protein extract. The three different protein fractions were separated by 1D SDS-PAGE followed by liquid chromatography mass spectrometry (LC-MS/MS) analysis of 12 gel bands digested with trypsin. This resulted in the collection of a total of 55,441 spectra, 14,342, 21,472 and 19,627 MS/MS spectra, respectively, obtained from the enriched bacterial fraction (EBF), faecal proteome fraction (FPF) and whole proteome fraction (WPF). The spectra were searched against the NCBI database (Supplementary data C4–1) and imported in Scaffold, which resulted in the identification of 645 proteins in 522 clusters ($\geq$1 peptide, protein threshold 99.0%, protein false discovery rate (FDR) 0.2%, peptide threshold 95.0% and peptide FDR 0.02%). With the same parameters but a minimum of two peptides, still 424 proteins were identified and grouped in 302 clusters. The largest number of proteins was identified in the faecal protein fraction, followed by the whole proteome fraction and the enriched bacterial fraction (Table 6.1 and Supplementary data C4–2). From the 424 proteins identified, 94 proteins were common between the three extraction methods (Supplementary data C4-3A), 11 proteins were only detected in the EBP (Supplementary data C4-3B), 122 proteins were only in the FPF (supplementary data C4-3C) and 101 proteins were only detected in the WPF (Supplementary data C4-3D). The FPF and WPF share 73 proteins and 11 and 12 proteins, respectively, with the EBF (Fig. 6.1).

6.3.2   Unipept Phylogenetic Analysis

The metaproteomic data generated were also used to determine the taxonomic origin of peptide identifications, with the aim of assessing any potential biases exerted by the different extraction methods. In Unipept [27], the lowest common ancestor approach is used and we applied a filter of a minimum of four different taxon-specific

Table 6.1 Protein quantitation. Proteins are identified by a minimum 2 peptides, a protein threshold of 99% and a peptide threshold 95%

Extraction method	Total number of spectra	# of identified spectra	% IDs	# of identified proteins
Enriched bacterial fraction	14,342	1612	11.0	128
Faecal proteome fraction	21,472	2049	9.5	300
Whole proteome fraction	19,627	2117	11.0	280

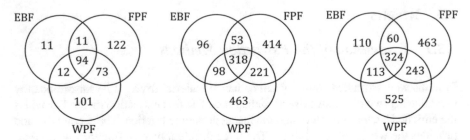

Fig. 6.1 Summary of the results of the different fractionation methods in terms of the number of identified proteins, the unique peptides and MS/MS spectra each originating from the different protein fractions. Venn diagrams were generated in Scaffold for proteins identified by a minimum of 2 peptides per protein, a peptide threshold of 95% and a protein threshold of 99%

peptides to determine if the respective taxon is present or not. In general, the number of taxon-specific peptides is 2720, 3307 and 3274 for the EBF, FPF and WPF, respectively (supplementary data C4–4). The highest abundance, measured by the summed spectral counts correlated to the bacteria is found in the WPF. Surprisingly, a higher number of taxon-specific peptides belonging to eukaryotes were found in the EBF (Fig. 6.2a). Moreover, the bacterial composition in the EBF seems to be biased; especially the phylum *Bacteroidetes* is overrepresented, while less taxon-specific peptides from the *Firmicutes* were observed (Fig. 6.2b), illustrating that not all bacteria are equally represented in the samples from the different methods used. At species-specific peptide level, the data is biased especially for *Bacteroides uniformis* but also for, although less pronounced, *Bacteroides vulgatus* and *Bacteroides dorei* (Fig. 6.2c). In the *Firmicutes*, less taxon-specific peptides were found in different families such as the *Ruminococcaceae* with *F. prausnitzii* and *Ruminococcus bromii* as main representatives and the family *Lachnospiraceae* with *Roseburia inulinivorans* as main representative (Fig. 6.2d). The results for the FPF and the WPF appear more similar than the EBF fraction.

6.3.3   Gene Ontology Analysis of the Different Extraction Methods

The identified proteins were further analysed using the Blast2GO program, which provides for each protein sequence an annotation with GO terminology to categorise the proteins according to their biological process, molecular function and cellular component (Fig. 6.3 and Supplementary data C4–5). Identified proteins appear to be involved in diverse biological processes. We obtained a total of 11, 14 and 13 subgroups for the EBF, FPF and WPF, respectively, as shown in Fig. 6.3a. The majority of identified proteins exhibit metabolic processes. The proteins in the WPF and FPF represent similar biological processes; the EBF covers less biological

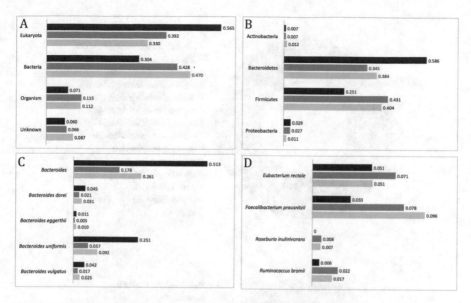

Fig. 6.2 Relative abundance of different taxonomy levels. The relative abundance of peptides from bacteria and eukaryotes (the detected taxon-specific peptides corrected for the total detected peptides) is visualised in panel (**a**) for the EBF (black), FPF (dark grey) and WPF (light grey). The relative abundance of the bacteria at phylum level, the detected taxon-specific peptides corrected for the detected bacterial peptides (panel **b**), the relative abundance of the genus *Bacteroides* (panel **c**) and the detected species in the order *Clostridiales* (panel **d**) are shown

processes. By analysing the proteins at GO level 3 in the category molecular function, we obtained a total of 7, 13 and 12 subgroups for the EBF, FPF and WPF (Fig. 6.3b). Here again the EBF is less rich in molecular functions, while FPF and WPF are more similar. The cellular localisation distribution appears to be quite similar in all three fractions (Fig. 6.3c), except for the remarkable contradiction that there are less GOs representing the cellular part and more from the extracellular region detected in the EBF.

6.3.4   Metaproteomic Data from One Patient with CF and Her Unaffected Sibling

Overall, it was concluded from the previous experiments that the EBF fraction was not well representing the bacterial diversity as assessed in the other fractions. The workload for additional centrifugation steps that were performed to obtain this fraction did not pay off. Therefore, we decided to evaluate the whole proteome extraction method in more detailed in a preliminary assessment of differences in the microbiota from patients with CF. Therefore, faecal samples from one patient with

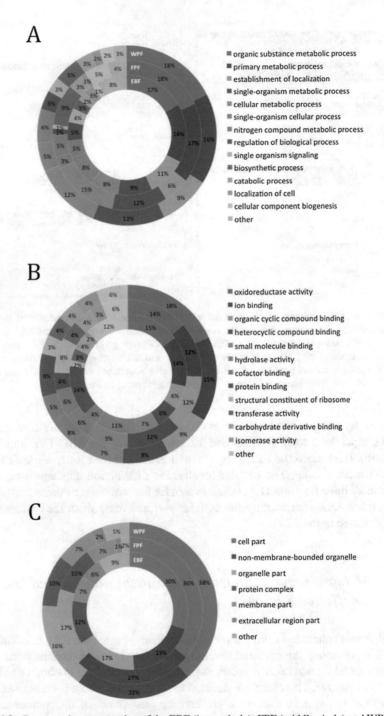

Fig. 6.3 Gene ontology annotation of the EBF (inner circle), FPF (middle circle) and WPF (outer circle). Each chart represents the distribution of the GO annotations into the major functional categories for each division, biological process (**a**), molecular function (**b**) and cellular component (**c**)

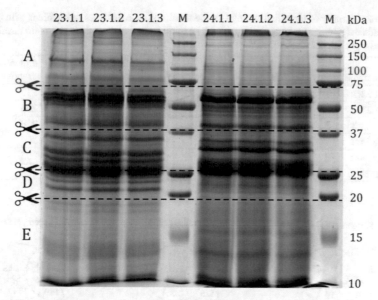

Fig. 6.4 1D-SDS-PAGE gel showing the protein pattern from the marker (M) and from faecal samples of patient 23 (Left) and sibling 24 (Right) on time point one (three technical replicates). Estimated molecular weights are indicated on the right. Dotted lines indicate where the gel was cut to obtain five fractions (**a–e**). Band patterns of the faecal metaproteome visualised a good reproducibility of the extraction protocol

CF and her unaffected sibling were analysed. A whole protein extract was prepared from 0.5 g faecal sample of a patient (labelled 23) (three replicates) and its sibling 24 (three replicates) collected approximately at the same time point. The protein extracts were separated by 1D SDS-PAGE and five gel bands (A-E) were digested with trypsin (Fig. 6.4) followed by liquid chromatography mass spectrometry (LC-MS/MS) analysis.

This resulted in the collection of 65,299 MS/MS spectra from patient and sibling in total. Of these MS/MS spectra, 25,054 led to the identification of 562 proteins at least represented by two different peptides. The number of proteins is more or less equally distributed between the three replicates of the patient with CF and the three replicates of the unaffected sibling (Table 6.2). Slightly more proteins were identified in the sample of the unaffected sibling, on average 301 proteins in contrast to an average of 225 proteins identified in the sample of the patient with CF.

The differences between the number of MS/MS spectra, unique peptides and proteins of the three technical replicates are shown in Fig. 6.5. For the three technical replicates from the patient, we found 156 proteins (51% of all proteins identified) common to all three replicates. (Fig. 6.5 upper panel left). Protein overlap exists, 51%, between all replicates from the same starting amount of faecal sample. Peptide overlap is 49% and MS/MS overlap is 46%. We also assessed the distribution of proteins identified among the three technical replicates of sibling 24. A total of 224 proteins (57% of all proteins identified) were common to all three samples

Table 6.2 Number of identified proteins of the three replicate analyses of a patient with CF and her unaffected sibling. Proteins are identified by a minimum of two peptides, a protein threshold of 99% and a peptide threshold of 95%

Sample name	Replica	Total number of spectra	# identified spectra	% IDs	# identified proteins
Patient 23r1	1	10,510	2252	21	232
Patient 23r2	2	10,000	2045	20	223
Patient 23r3	3	11,152	2181	20	220
Sibling 24r1	1	11,085	2516	23	304
Sibling 24r2	2	11,277	2542	23	295
Sibling 24r3	3	11,275	2705	24	304

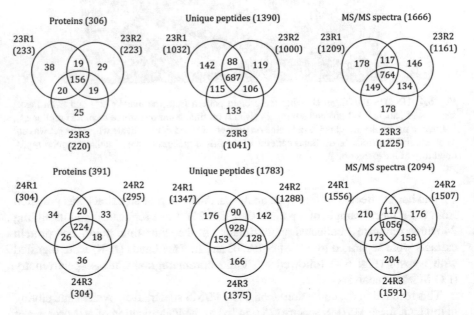

Fig. 6.5 LC-MS/MS identified proteins (left), the unique peptides (middle) and MS/MS spectra (right) of the three replicates from patient 23 (upper panel) and sibling 24 (lower panel). Venn diagrams were generated in Scaffold for proteins identified by a minimum of two peptides per protein, a peptide threshold of 95% and a protein threshold of 99%

(Fig. 6.5 lower panel). As found with the technical replicates of patient 23, a significant portion of the proteins identified, 57%, was common to all three replicates from sibling 24 and was slightly more than the overlap between the technical replicates from patient 23. Peptide overlap is 52% and MS/MS overlap is 50%.

Finally, we assessed the distribution of proteins, peptides and MS/MS spectra identified in the three replicates of the patient and the three replicates of the sibling (Fig. 6.6). Of the 562 proteins detected, only 135 proteins (24% of all proteins identified) were found in common (Supplementary data C4-6A). 171 proteins were detected

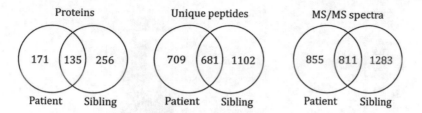

Fig. 6.6 LC-MS/MS identified proteins (left), the unique peptides (middle) and unique MS/MS spectra (right) in the patient with CF and in her sibling. Venn diagrams were generated in Scaffold for proteins identified by a minimum of two peptides per protein, a peptide threshold of 95% and a protein threshold of 99%

exclusively in the sample of the patient (Supplementary data C4-6B) and 256 proteins exclusively in the unaffected sibling (Supplementary data C4-6C). Peptide overlap is 27% and MS/MS spectra overlap is 28% between patient 23 and sibling 24.

6.3.5   Phylogenetic Analysis of the Data from a Patient with CF and Her Corresponding Unaffected Sibling

We further investigated the differences between the three replicates of the patient with CF or her unaffected sibling at taxonomy level. Only taxa with an average count of minimal four distinct peptides in one of the samples were considered for further analysis. The phylogenetic analysis reveals 35 taxon-specific levels in the patient and 52 in the sibling (Supplementary data C4–7). The peptide counts for the phylum taxa were normalised against the total number of bacterial peptides detected in the sample, resulting in the relative abundance of bacterial taxa. Unipept analysis showed the representatives of the four main phyla previously found within the human gut [32]. *Firmicutes*, *Bacteroidetes*, *Actinobacteria* and *Proteobacteria* were present in all replicates of both the patient and the sibling (Fig. 6.7). For the *Actinobacteria*, *Firmicutes* and the *Proteobacteria*, a statistical difference was found between the samples originating from the patient with CF and those from her sibling. The variation between the three replicates was higher in the three replicates from the patients than in the three replicates from the sibling.

The most differences between the sample of patient and her sibling were noticed within the phylum *Firmicutes* (Fig. 6.8). The phylum *Firmicutes*, class *Clostridia*, order *Clostridiales*, family *Eubacteriaceae*, genus *Eubacterium*, species *Eubacterium rectale*, family *Ruminococcaceae*, genus *Faecalibacterium*, species *Faecalibacterium* prausnitzii, genus *Ruminococcus* and species *Ruminococcus bromii* have statistically more taxon-specific peptides detected in the sibling. The species *Eubacterium rectale*, *Faecalibacterium prausnitzii* and *Ruminococcus bromii* are uniquely detected in the sibling. On the other hand, the family *Clostridiaceae*, genus *Clostridium*, genus *Blautia* and species *Blautia gnava* [33] have more taxon-specific peptides detected in the patient.

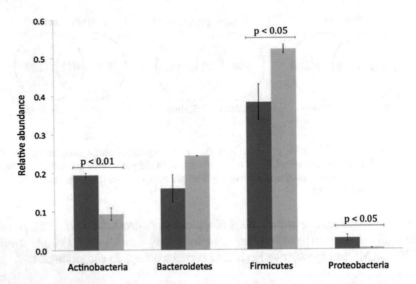

Fig. 6.7 Unipept results showing taxonomic distribution of the detected phyla. The taxa of the patient are represented by dark grey columns and the sibling by light grey columns. The relative abundances present the peptide counts for the phylum of subject *i* divided by the peptide counts for the Kingdom *Bacteria*. Student's t-test was used to calculate p-values

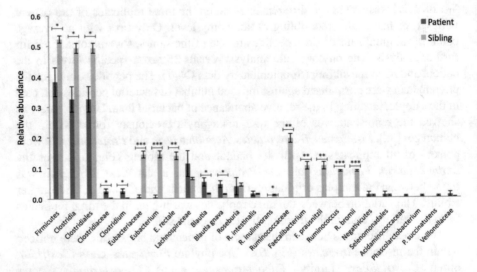

Fig. 6.8 Unipept results showing taxonomic distribution within the phylum *Firmicutes*. The taxa of the patient are represented by dark grey columns, those of the sibling by light grey columns. The relative abundances are calculated as the peptide counts for the taxa of subject *i* divided by the total peptide counts for the Kingdom *Bacteria*. Data represented as means ± SD and the Student's t-test was used to calculate p-values, $*p < 0.05$, $**p < 0.01$ and $***p < 0.001$

6.3.5.1   Gene Ontology

The detected proteins from every replicate were analysed separately using the Blast2GO program, which provides for each protein sequence an annotation with GO terminology to categorise the proteins in biological process, molecular function and cellular component (Supplementary data C4–8). The most abundant GO terms of the three replicates are similar, the same GO terms are recovered in each replicate and the percentages detected in the three replicates are also quite similar (Fig. 6.9).

In addition, a GO enrichment analysis was performed. Therefore, the data from the three replicates of the patient with CF were merged and compared to the merged data of the three replicates of the unaffected sibling (Supplementary data C4–9). The GO data reduced to the most specific GO term annotation is shown in Fig. 6.10. Several biological processes were underrepresented in the patient, such as the regulation of translational elongation and carbohydrate transport. Only one molecular function, protein binding, was enriched in the patient. In the sibling on the other hand, the translation elongation factor activity, GTP binding and GTPase activity were enriched. Regarding the cellular localisation, some GO terms were enriched in the patient, particularly the extracellular space. These proteins correspond to secreted host proteins such as immunoglobulin, trefoil factor 2, zymogen granule membrane proteins and so on. In the healthy person, cellular localisation of the ribosome was enriched.

The KEGG analysis revealed that the proteins identified via the metaproteomic approach are representing proteins from 32 different pathways. Twelve pathways remained after filtering on a minimum presence of two enzymes in at least one of the replicates (Table 6.3). Most of the pathways were detected with minimal two peptides in all three replicates, except for the pentose phosphate metabolism, methane metabolism and the starch and sucrose metabolism.

In brief, our data suggest that the faecal microbiota composition and a variety of functions are reproducible between the three replicates of the patient or sibling. There are, however, differences between the results of the samples of the patient with CF and that of her unaffected sibling. This suggests that our method for investigating the faecal microbiota, both taxonomical and functional, is useful for large-scale metaproteomic studies to determine the differences between patients with CF and their unaffected sibling.

6.4   Discussion

The human gut harbours a complex community of microbes that profoundly influences the health and well-being of their host in many aspects including growth and development [34]. Cystic fibrosis alters the intestinal environment and therefore the uptake of nutrients. Besides, long-term antibiotic treatments change the intestinal microbiota and to investigate such dysbiosis, we developed a metaproteomic pipeline to investigate via a non-invasive manner the proteins present in faecal samples

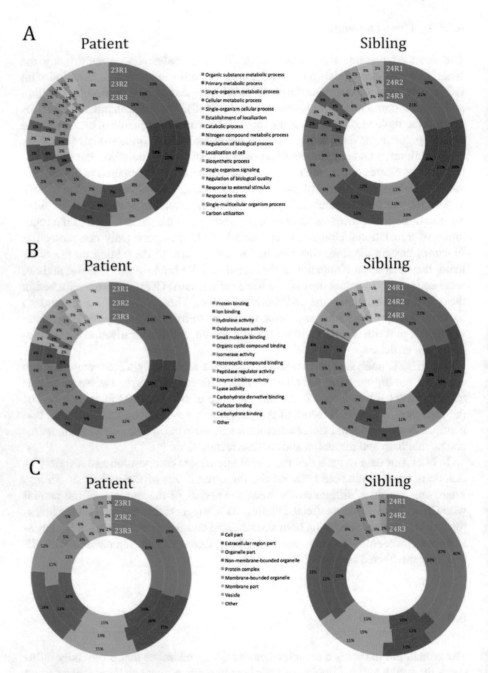

Fig. 6.9 Percentage representation of Gene ontology (GO) level 3 mappings for the three replicates from the patient with CF (left) and the three replicates of her unaffected sibling (right). Panel
(**a**) are the biological processes, panel (**b**) the molecular functions and panel (**c**) the cellular
components

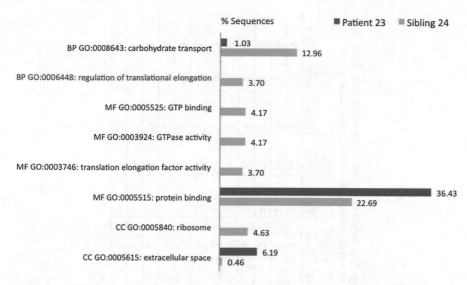

Fig. 6.10 Go enrichment analysis from Blast2GO. GO terms identified with enrichment analysis comparing the merged data of the proteins identified in the three replicates of the patient with CF (black) against the merged data of the proteins identified in the three replicates of the unaffected sibling (grey)

of patients with cystic fibrosis and compare them with their unaffected sibling. First, we compared two extraction methods, which resulted in three fractions, the enriched bacterial fraction (EBF), the faecal proteome fraction (FPF) and the whole proteome fraction (WPF). Next, we evaluated the optimised method on three technical replicates originating from a faecal sample from a patient with CF and three technical replicates from the faecal sample from her unaffected sibling.

6.4.1   Evaluation of the Two Extraction Methods

Several aspects of an LC-MS workflow, such as protein extraction, sample and data processing, bioinformatics and experimental set-up, influence quantitative metaproteomic analyses. In this study, we concentrated on one of these parameters, namely, the protein extraction from faecal samples stored in the freezer ($-80\ ^{\circ}$C). It is known that more than half of the cells found in faecal samples are in a non-viable or heavily damaged state [35] and for faecal samples stored in the freezer, this is even worse [36]. In an ideal situation, fresh samples should be used for metaproteomic analyses, but for practical reasons, mainly in terms of sample collection at different time points, this is unfeasible.

The spectral counting label-free method used in this study is an example of a quick and effective method for detecting large differences in proteins extracted via different methods [37]. All three protein extracts allowed for detecting both

Table 6.3 KEGG pathways detected in patients with CF and their corresponding unaffected siblings. The number of sequences detected in the pathway (Seqs in pathway) and the number of different enzymes (# Enzymes) detected with a minimum of two different enzymes in one of the samples are displayed

Kegg pathways	23R1		23R2		23R3		24R1		24R2		24R3	
	Seqs	Enzs	Seqs	Enzs	Seqs	Enzs	Seqs	Enzs	Seqs	Enzs	Seqs	Enzs
Carbon fixation in photosynthetic organisms	7	3	10	3	8	3	15	4	12	3	13	5
Glycolysis/gluconeogenesis	7	3	9	3	7	2	17	4	13	3	13	5
Fructose and mannose metabolism	6	3	7	3	6	3	12	4	10	4	11	5
Pyruvate metabolism	5	3	7	3	6	3	10	3	8	2	9	3
Pentose and glucuronate interconversions	5	3	5	4	6	3	10	2	9	2	11	2
Citrate cycle (TCA cycle)	4	2	5	2	5	2	9	2	7	1	8	2
Carbon fixation pathways in prokaryotes	2	2	5	3	2	2	3	3	1	1	3	3
Pentose phosphate pathway	2	1	1	1	1	1	3	2	1	1	3	3
Methane metabolism	2	2	3	2	1	1	3	2	2	1	3	3
Glyoxylate and dicarboxylate metabolism	2	2	4	2	2	2	1	1	0	0	1	1
Starch and sucrose metabolism	1	1	2	2	2	2	2	1	0	0	1	1
Tyrosine metabolism	1	2	1	2	1	2	0	0	0	0	0	0

eukaryotic and bacterial proteins in the faecal sample from a healthy individual. Of the 20 most abundant protein clusters present in all three fractions, half are of eukaryotic origin (human proteins) and the other half are bacterial proteins (Supplementary data C4-3A). Among the human proteins are, as expected, different digestive enzymes, keratins and immunoglobulins. In addition, we found high levels of protocadherin LKC, a calcium dependent mediator of cell-cell adhesion that associates with the mucosal actin cytoskeleton and of which the gene is widely expressed in colon tissues [38]. Of the bacterial proteins, the cluster of glutamate dehydrogenase (GDH) was found to be abundant in our data in accordance with a previous metaproteomic study on healthy subjects [24]. These authors also mentioned a high level of protein functional redundancy in the intestinal tract. We could identify two clusters of GDH referring to the *Firmicutes* and *Bacteroidetes*, respectively. The glycolytic enzyme glyceraldehyde-3-phosphate dehydrogenase (GAPDH) was identified in similar clusters. Furthermore, there were many ABC sugar transporters identified which is also in-line with the results from previous studies [24]. The abundance of these sugar transporters underlines the importance of sugar transport and metabolism in the intestinal microbiota. In addition many hypothetical proteins (five clusters from the top 20) are detected, which was also reported in the study of Verberkmoes et al., where 20% of the identified proteins were classified as "hypothetical proteins" [22].

Only eleven proteins were uniquely detected in EBF (Supplementary data C4-3B). In the faecal protein fraction (FPF), more proteins were detected that were unique to this fraction (Supplementary data C4-3C). The most abundant proteins found uniquely in the WPF (Supplementary data C4-3D) are all bacterial: six from the genus *Bacteroides* and four from the phylum *Firmicutes*. Curiously, we demonstrate that the WPF yields the highest number of identified spectra. In addition, a higher number of peptides were identified in the WPF, although they were originating from a smaller number of proteins. If we look at the results of the Unipept analysis, we remarkably found a higher number of taxon-specific peptides belonging to eukaryotes in the EBF whereas the highest number of spectral counts correlated to the bacteria is found in the WPF.

As we compared the extraction efficiency of proteins from a sample for which the bacterial cells were enriched with a sample where no prefractionation was used, we were able to evaluate the throughput of our extraction protocols. For the enrichment of the bacterial cells, a one-step enrichment was used. We realise that this method is less efficient than previously described enrichment procedures that used four differential centrifugation steps [39] or Nycodenz density gradient centrifugation [23]. However, we have demonstrated that a one-step enrichment of the bacterial fraction in comparison with the whole protein fraction results in poorly repeatable results, i.e. identification of different bacterial proteins from the same sample. This affects the compositional and functional interpretation of the microbial community. In addition to the differences in bacterial composition, the purpose to enrich the bacterial fraction from frozen faecal samples fails since actually more host proteins are detected. This was also noticed by Kolmeder et al. [40], who showed that metaproteomics using unpurified faeces reveals more microbial (85%) proteins than a differential centrifugation step (70%).

6.4.2   Replicate Analysis of Faecal Sample from a Patient with CF and Her Unaffected Sibling

Based on these data, we decided to use the whole protein fraction to set up an exploratory experiment to investigate the metaproteomic differences in three replicates from frozen faecal samples from a patient with CF and her unaffected sibling. The used protocol was slightly adapted as only 0.5 gram of faecal sample was used, the centrifugation steps were prolonged, and instead of twelve, six bands were cut from the gel.

At the protein level, the three technical replicates show similar results in number of detected proteins, unique peptides and MS/MS spectra within the three patient analyses and within the three sibling analyses. Most MS/MS spectra, peptides and proteins are detected in all three replicates, but additional identifications were made with each added technical replicate. Up to six technical replicates are required for optimal identification for proteomic samples, and a more complex sample, such as metaproteomic samples, may require additional technical replicates for optimal identification [41]. Evidently, this is hard to achieve for metaproteomic analyses from faecal samples, especially for the patients, where the obtained faecal matter is less than 1 gram. Besides, it should be realised that for studying the faecal microbiota of patients with CF, multiple patient-sibling pairs should be included resulting in an exponential increasing analysis time. There is a considerable variation in the identified proteins in the replicate samples. This is most likely due to the undersampling nature of data-dependent LC-MS-MS analysis which can be tackled by more extended two-dimensional chromatography. It should be stressed also that the current Orbitrap mass spectrometers have a dramatically improved duty cycle compared to our FTICR-MS systems and allow a deeper metaproteomic coverage [42].

We commonly identified 135 proteins in the samples obtained from the patient and its sibling. The top ten protein clusters of these proteins in common are all proteins of human origin (Supplementary data C4-6A). The most abundant protein clusters contain immunoglobulin-related proteins (cluster of immunoglobulin kappa light chain, cluster of immunoglobulin lambda 2 light chain, cluster of IgGFc-binding protein precursor) and some digestive enzymes, such as the cluster of aminopeptidase N precursor, the cluster of pancreatic alpha-amylase and the cluster of maltase-glucoamylase. Of the ten most abundant proteins, exclusively present in the patient with CF and not detected in the unaffected sibling (Supplementary data C4-6B), six proteins are of porcine origin (*Sus scrofa*). Patients with pancreatic insufficiency take high-dose pancreatic enzyme replacement therapy (PERT) to compensation for the lack of pancreatic enzymes. Of the ten most abundant proteins exclusively present in the unaffected sibling, only two human proteins were detected (Supplementary data C4-6C), elastase 3A and human procarboxypeptidase B; both proteins are secreted by the pancreas, which corresponds to the fact that the patient with CF in this study has pancreatic insufficiency. The other seven proteins were bacterial and mainly consist of ABC transporters (*F. prausnitzii*, *Eubacterium rectale* and

Bifidobacterium adolescentis) and hypothetical proteins (*Ruminococcus* sp., *E. rectale* and *Bacteroides*).

The taxonomic analysis by Unipept shows that the results of the unaffected sibling sample are similar to the defined healthy microbiota which is motivated by a consensus between different consortiums with an acceptation of 60% *Firmicutes*, 25% *Bacteroidetes* and 15% minor phyla [43]. In this patient-sibling pair, significantly more spectral counts are detected in the patient with CF originating from proteins from the phyla *Actinobacteria* and *Proteobacteria*. The enrichment of the *Proteobacteria* and more specific the enterobacterial counts has been previously noticed in patients with CF [44–47] and Crohn's disease [47, 48] and has been confirmed in an extended patient/sibling comparative analysis described by us elsewhere [49]. The dysbiosis may be due to the frequent administration of antibiotics [50–53], the persistent diarrhoea and alterations of GI motility and/or the bacterial overgrowth in the small intestine. High levels of *Actinobacteria*, mainly *Bifidobacterium longum*, were present in the patient with CF. This, however, can be explained by the patient's age. The child with CF is only 2 years, and *Bifidobacterium longum* is one of the most prevalent species in young infants [54]. Nevertheless, it proves that the followed strategy is able to pick up differences in the faecal microbiota.

6.5   Conclusion

A metaproteomic approach was developed to detect the differences in faecal proteins between patients with CF and their unaffected sibling. We showed that the whole protein fraction extraction method provides us more information compared to a method including a bacterial cell enrichment step. Our preliminary results confirm and extend previous observations based on denaturing gradient gel electrophoresis fingerprints and quantitative PCR experiments and paved the way to a study of a larger cohort of patients/sample pairs. We have developed a promising approach to characterise and understand the dysbiosis of the gut microbiota in patients with CF. Briefly, the whole protein fraction method is used to extract the proteins from the faecal samples. Subsequently, the extracted proteins are separated by 1D SDS-PAGE, and five gel bands were digested with trypsin followed by liquid chromatography mass spectrometry (LC-MS/MS) analysis. By integrating our shotgun metaproteomics with statistical and computational analyses, our approach should be applicable to give functional insights into the metaproteomes of complex ecosystems such as the human microbiota.

We observed that a wealth of information is available in the faecal metaproteome data. Not only the microbial proteins are of interest; also the host proteome provides details of the intestinal status. Although there are still limitations, in particular the low coverage of peptide identifications and dynamic range, we obtained a workflow that is less biased to certain bacterial species and provides information on the functionality of the intestinal microbiota.

Acknowledgements This research was supported by grant G.0638.10 from Research Foundation Flanders (FWO). PD acknowledges the support of Ghent University (MRP Bioinformatics: from nucleotides to networks). The authors thank Dr. Kris Moreel for the generous help with the LC-MS/MS analyses on the FT-ICR-MS.

Supplementary Data

Supplementary files can be downloaded from http://users.ugent.be/~bdevrees/.

References

1. Dethlefsen L, Eckburg PB, Bik EM et al (2006) Assembly of the human intestinal microbiota. Trends Ecol Evol 21:517–523
2. Gill SR, Pop M, DeBoy RT et al (2006) Metagenomic analysis of the human distal gut microbiome. Science 312:1355–1359
3. Cummings JH, Macfarlane GT (1997) Role of intestinal bacteria in nutrient metabolism. JPEN J Parenter Enteral Nutr 21:357–365
4. Guarner F, Malagelada JR (2003) Gut flora in health and disease. Lancet 361:512–519
5. Sekirov I, Russell SL, Antunes LC et al (2010) Gut microbiota in health and disease. Physiol Rev 90:859–904
6. Pryde SE, Duncan SH, Hold GL et al (2002) The microbiology of butyrate formation in the human colon. FEMS Microbiol Lett 217:133–139
7. Ubeda C, Pamer EG (2012) Antibiotics, microbiota, and immune defense. Trends Immunol 33:459
8. Mortensen PB, Clausen MR (1996) Short-chain fatty acids in the human colon: relation to gastrointestinal health and disease. Scand J Gastroenterol Suppl 216:132–148
9. Barcenilla A, Pryde SE, Martin JC et al (2000) Phylogenetic relationships of butyrate-producing bacteria from the human gut. Appl Environ Microbiol 66:1654–1661
10. Duytschaever G, Huys G, Bekaert M et al (2013) Dysbiosis of bifidobacteria and Clostridium cluster XIVa in the cystic fibrosis fecal microbiota. J Cyst Fibros 12:206–215
11. Tannock GW (2008) The search for disease-associated compositional shifts in bowel bacterial communities of humans. Trends Microbiol 16:488–495
12. Larsen N, Vogensen FK, van den Berg FW et al (2010) Gut microbiota in human adults with type 2 diabetes differs from non-diabetic adults. PLoS One 5:e9085
13. van Tongeren SP, Slaets JP, Harmsen HJ et al (2005) Fecal microbiota composition and frailty. Appl Environ Microbiol 71:6438–6442
14. Balamurugan R, Rajendiran E, George S et al (2008) Real-time polymerase chain reaction quantification of specific butyrate-producing bacteria, Desulfovibrio and *Enterococcus faecalis* in the feces of patients with colorectal cancer. J Gastroenterol Hepatol 23:1298–1303
15. Davies JC, Bilton D (2009) Bugs, biofilms, and resistance in cystic fibrosis. Respir Care 54:628–640
16. Wilschanski M, Durie PR (2007) Patterns of GI disease in adulthood associated with mutations in the CFTR gene. Gut 56:1153–1163
17. O'Brien S, Mulcahy H, Fenlon H et al (1993) Intestinal bile acid malabsorption in cystic fibrosis. Gut 34:1137–1141
18. Duytschaever G, Huys G, Bekaert M et al (2011) Cross-sectional and longitudinal comparisons of the predominant fecal microbiota compositions of a group of pediatric patients with cystic fibrosis and their healthy siblings. Appl Environ Microbiol 77:8015–8024

19. Bruzzese E, Raia V, Gaudiello G et al (2004) Intestinal inflammation is a frequent feature of cystic fibrosis and is reduced by probiotic administration. Aliment Pharmacol Ther 20:813–819
20. Hawrelak JA, Myers SP (2004) The causes of intestinal dysbiosis: a review. Altern Med Rev 9:180–197
21. Modolell I, Guarner L, Malagelada JR (2002) Digestive system involvement in cystic fibrosis. Pancreatology 2:12–16
22. Verberkmoes NC, Russell AL, Shah M et al (2009) Shotgun metaproteomics of the human distal gut microbiota. ISME J 3:179–189
23. Rooijers K, Kolmeder C, Juste C et al (2011) An iterative workflow for mining the human intestinal metaproteome. BMC Genomics 12:6
24. Kolmeder CA, de Been M, Nikkila J et al (2012) Comparative metaproteomics and diversity analysis of human intestinal microbiota testifies for its temporal stability and expression of core functions. PLoS One 7:e29913
25. Keller A, Nesvizhskii AI, Kolker E et al (2002) Empirical statistical model to estimate the accuracy of peptide identifications made by MS/MS and database search. Anal Chem 74:5383–5392
26. Nesvizhskii AI, Keller A, Kolker E et al (2003) A statistical model for identifying proteins by tandem mass spectrometry. Anal Chem 75:4646–4658
27. Mesuere B, Devreese B, Debyser G et al (2012) Unipept: tryptic peptide-based biodiversity analysis of metaproteome samples. J Proteome Res 11:5773–5780
28. Conesa A, Gotz S, Garcia-Gomez JM et al (2005) Blast2GO: a universal tool for annotation, visualization and analysis in functional genomics research. Bioinformatics 21:3674–3676
29. Quevillon E, Silventoinen V, Pillai S et al (2005) InterProScan: protein domains identifier. Nucleic Acids Res 33:W116–W120
30. Myhre S, Tveit H, Mollestad T et al (2006) Additional gene ontology structure for improved biological reasoning. Bioinformatics 22:2020–2027
31. Kanehisa M, Goto S, Sato Y et al (2012) KEGG for integration and interpretation of large-scale molecular data sets. Nucleic Acids Res 40:D109–D114
32. Detlefsen L, McFall-Ngai M, Relman DA (2007) An ecological and evolutionary perspective on human-microbe mutualism and disease. Nature 449:811–818
33. Liu C, Finegold SM, Song Y et al (2008) Reclassification of Clostridium coccoides, Ruminococcus hansenii, Ruminococcus hydrogenotrophicus, Ruminococcus luti, Ruminococcus productus and Ruminococcus schinkii as Blautia coccoides gen. nov., comb. nov., Blautia hansenii comb. nov., Blautia hydrogenotrophica comb. nov., Blautia luti comb. nov., Blautia producta comb. nov., Blautia schinkii comb. nov. and description of Blautia wexlerae sp. nov., isolated from human faeces. Int J Syst Evol Microbiol 58:1896–1902
34. Gerritsen J, Smidt H, Rijkers GT et al (2011) Intestinal microbiota in human health and disease: the impact of probiotics. Genes Nutr 6:209–240
35. Ben-Amor K, Heilig H, Smidt H et al (2005) Genetic diversity of viable, injured, and dead fecal bacteria assessed by fluorescence-activated cell sorting and 16S rRNA gene analysis. Appl Environ Microbiol 71:4679–4689
36. Bahl MI, Bergstrom A, Licht TR (2012) Freezing fecal samples prior to DNA extraction affects the Firmicutes to Bacteroidetes ratio determined by downstream quantitative PCR analysis. FEMS Microbiol Lett 329:193–197
37. Neilson KA, Ali NA, Muralidharan S et al (2011) Less label, more free: approaches in label-free quantitative mass spectrometry. Proteomics 11:535–553
38. Okazaki N, Takahashi N, Kojima S et al (2002) Protocadherin LKC, a new candidate for a tumor suppressor of colon and liver cancers, its association with contact inhibition of cell proliferation. Carcinogenesis 23:1139–1148
39. Apajalahti JH, Sarkilahti LK, Maki BR et al (1998) Effective recovery of bacterial DNA and percent-guanine-plus-cytosine-based analysis of community structure in the gastrointestinal tract of broiler chickens. Appl Environ Microbiol 64:4084–4088
40. Kolmeder CA, de Vos WM (2014) Metaproteomics of our microbiome - developing insight in function and activity in man and model systems. J Proteome 97:3–16

41. Ham BM, Yang F, Jayachandran H et al (2008) The influence of sample preparation and replicate analyses on HeLa cell phosphoproteome coverage. J Proteome Res 7:2215–2221
42. Zhang X, Chen W, Ning Z et al (2017) Deep Metaproteomics approach for the study of human microbiomes. Anal Chem 89:9407–9415
43. Lozupone CA, Stombaugh JI, Gordon JI et al (2012) Diversity, stability and resilience of the human gut microbiota. Nature 489:220–230
44. Duytschaever G, Huys G, Boulanger L et al (2013) Amoxicillin-clavulanic acid resistance in fecal Enterobacteriaceae from patients with cystic fibrosis and healthy siblings. J Cyst Fibros 12:780
45. Schippa S, Iebba V, Santangelo F et al (2013) Cystic fibrosis transmembrane conductance regulator (CFTR) allelic variants relate to shifts in faecal microbiota of cystic fibrosis patients. PLoS One 8:e61176
46. Del Campo R, Garriga M, Perez-Aragon A et al (2014) Improvement of digestive health and reduction in proteobacterial populations in the gut microbiota of cystic fibrosis patients using a *Lactobacillus reuteri* probiotic preparation: a double blind prospective study. J Cyst Fibros 13:716–722
47. Sokol H, Seksik P, Furet JP et al (2009) Low counts of Faecalibacterium prausnitzii in colitis microbiota. Inflamm Bowel Dis 15:1183–1189
48. Willing BP, Dicksved J, Halfvarson J et al (2010) A pyrosequencing study in twins shows that gastrointestinal microbial profiles vary with inflammatory bowel disease phenotypes. Gastroenterology 139:1844–1854 e1841
49. Debyser G, Mesuere B, Clement L et al (2016) Faecal proteomics: a tool to investigate dysbiosis and inflammation in patients with cystic fibrosis. J Cyst Fibros 15:242–250
50. Ladirat SE, Schols HA, Nauta A et al (2013) High-throughput analysis of the impact of antibiotics on the human intestinal microbiota composition. J Microbiol Methods 92:387–397
51. Jernberg C, Lofmark S, Edlund C et al (2007) Long-term ecological impacts of antibiotic administration on the human intestinal microbiota. ISME J 1:56–66
52. Jakobsson HE, Jernberg C, Andersson AF et al (2010) Short-term antibiotic treatment has differing long-term impacts on the human throat and gut microbiome. PLoS One 5:e9836
53. Perez-Cobas AE, Gosalbes MJ, Friedrichs A et al (2013) Gut microbiota disturbance during antibiotic therapy: a multi-omic approach. Gut 62:1591–1601
54. Turroni F, Peano C, Pass DA et al (2012) Diversity of Bifidobacteria within the infant gut microbiota. PLoS One 7:e36957

Chapter 7
Sample Preparation Focusing on Plant Omics

Rodrigo Moretto Galazzi, Jemmyson Romário de Jesus, and Marco Aurélio Zezzi Arruda

Abbreviations

AAM	Ammonium acetate/methanol
ADP	Adenosine diphosphate
ATP	Adenosine triphosphate
BIF	Banded iron formation
CE-ICP-MS	Capillary electrophoresis-inductively coupled plasma-mass spectrometry
DNA	Deoxyribonucleic acid
ESI-MS	Electrospray ionization mass spectrometry
ESI-FAIMS-IT-MS	Electrospray ionization-high-field asymmetric waveform ion mobility spectrometry-ion trap mass spectrometry
FRET	Fluorescence resonance energy transfer
FT-ICR MS	Fourier transform ion cyclotron resonance mass spectrometry
HPLC-DAD	High-performance liquid chromatography with diode array detector
HPLC-DAD-ESI-MS/MS	High-performance liquid chromatography with diode array detector coupled to electrospray ionization tandem mass spectrometry
HPLC-UV	High-performance liquid chromatography with ultraviolet detector
HRE	Heat reflux extraction
ICAT	Isotope-coded affinity
ICP-MS	Inductively coupled plasma mass spectrometry
IEF	Isoelectric focusing

R. M. Galazzi · J. R. de Jesus · M. A. Z. Arruda (✉)
Universidade Estadual de Campinas – Unicamp, Campinas, Brazil
e-mail: zezzi@iqm.unicamp.br

© Springer Nature Switzerland AG 2019
J.-L. Capelo-Martínez (ed.), *Emerging Sample Treatments in Proteomics*, Advances in Experimental Medicine and Biology 1073, https://doi.org/10.1007/978-3-030-12298-0_7

iTRAQ	Isobaric tags for relative and absolute quantification
JA	Jasmonic acid
LC	Liquid chromatography
LC-ESI-MS/MS	Liquid chromatography coupled to electrospray ionization tandem mass spectrometry
LC-ICP-MS	Liquid chromatography-inductively coupled plasma mass spectrometry
LC-MS/MS	Liquid chromatography tandem-mass spectrometry
LC-MS	Liquid chromatography mass spectrometry
MAE	Microwave-assisted extraction
MALDI-MS	Matrix-assisted laser desorption/ionization coupled mass spectrometry
ME	Maceration
MS	Mass spectrometry
MTBE	Methyl *tert*-butyl ether
MUDPIT	Multidimensional protein identification technology
NADP	Nicotinamide adenine dinucleotide phosphate
nanoESI-Q-TOF	Nano-electrospray ionization quadrupole time-of-flight
nanoSIMS	Nanoscale secondary ion mass spectrometry
NMR	Nuclear magnetic resonance
PARC	PEI-assisted RuBisCO cleanup
PEI	Polyethylenimine
PEG	Polyethylene glycol
pI	Point isoelectric
PM	Sodium phosphate/methanol
PR	Pathogenesis related
PTMs	Posttranslational modifications
RP-HPLC	Reversed-phase chromatography
RP-HPLC-UV-ESI-MS	Reversed-phase high-performance liquid chromatography-ultraviolet-electrospray ionization mass spectrometry
RuBisCO	Ribulose-1,5-bisphosphate carboxylase/oxygenase
SEC-UV	Size exclusion chromatography-ultraviolet
SDS-PAGE	Sodium dodecyl sulfate polyacrylamide gel electrophoresis
(TAP)-MS	Tandem affinity purification mass spectrometry
TCA	Trichloroacetic acid
UAE	Ultrasound-assisted extraction
UHPLC-DAD-ESI-MS/MS	Ultrahigh-performance liquid chromatography with diode array detector coupled to electrospray ionization tandem mass spectrometry
UHPLC-HR-MS	Ultrahigh-performance liquid chromatography coupled with high-resolution mass spectrometry
WM	Water/methanol

7.1 Introduction

Although considered as old as the analyses, sample preparation is always reinventing and adapting itself [2]. Nowadays, it can be updated with the advent of the omics science, inside the context of plant omics (and for other omics), once some necessities are required, due to a diversity of analytes, their complexities, quantities, and concentrations, thus requiring the modus operandi of sample preparation of customized strategies.

Once the analytical information is analyte dependent, then one of the key points in terms of sample preparation is to maintain the integrity of the analyte along the process. For example, as metallomics links to genomics and proteomics, it may come to mind similar sample preparation procedures, which could be applied to all areas. However, while proteins mostly present labile compounds (organic) in their composition, metalloproteins present both labile and stable (metals, metalloids) constituents. Only this fact makes sample preparation be different when the target is metallomics or proteomics [25, 67].

As noted, the challenge is huge in terms of sample preparation focusing on biomolecules, which meets those essential analytical characteristics, such as selectivity, robustness, precision, and accuracy. The challenge is great even considering plant science, which has been highlighting rapid progress in this field in plants, with emphasis on recent work in model plant species, subcellular organelles, and specific aspects of the plant life cycle such as signaling, reproduction, and stress physiology. If by one side plants are considered simpler structured organism compared to human due their low number of organs and tissue types, by the other side the complexity origin of the plants comes from the high number and variety of their species, >300,000 compared to 5000 in mammals [1].

However, these problems can be faced today, once we have sort of possibilities in terms of strategies and techniques. It is precisely within this scenario that we focus on this chapter, and a diversity of strategies, techniques, and procedures will be highlighted and commented.

7.2 Sample Preparation Focusing on Plant Proteomics: Challenges and Achievements

Plants represent the majority of the eukaryotic biomass of the planet and play an important role for the existence of living organisms, once they are the fundamental basis of human and animal nutrition. In addition, they are also used as invaluable resources for both renewable raw material and energy, and they synthesize a great variety of essential molecules (amino acids, vitamins, lipids, and secondary metabolites), including highly efficient pharmaceuticals. In this way, the plants can be considered as the most important species on our planet [33]. Another factor that justifies this statement is unique physiological properties exhibited by plant, which are not deduced from studies in other organisms. Such properties are realized notably in response to their biotic and abiotic environments [33].

However, the contamination of soils and water with metals or other toxic substances has created a major environmental problem, leading to considerable losses in plant productivity and hazardous health effects [33, 44]. For these reasons, understanding the unique functions and biological processes at the molecular level from plant is crucial to improve the security and productivity of cultivated areas within their social, political, and economic context. In this way, proteomics emerges as an important part of plant science.

Proteomics is the large-scale study of proteins, particularly their structures and functions. Proteins are vital parts of living organisms, as they are the main components of the physiological metabolic pathways of cells. Proteomics is an interdisciplinary domain formed on the research basis and development of the Human Genome Project, and it has also consolidated in scientific research of proteome exploration from the overall level of intracellular protein composition, structure, and its own unique activity patterns. Proteomics is an important component of functional genomics, and it is the large-scale functional analysis of proteins extracted from intact organisms, tissues, individual cells, or cell compartments, at defined time points during development or under specific conditions. With the avalanche of genomic information and improvements in analytical technology, proteomics has become increasingly important for the study of many different aspects of plant functions, since proteins serve as important components of major signaling and biochemical pathways.

Plant proteomics highlights the rapid progress in this field, and in this way, proteomic studies have focused on many different plant species in order to understand how the plants grow and interact with the environment as well as expand the information regarding the consequences of genetic modifications [1, 3, 4, 44].

Furthermore, we describe some specific structures found in plants and highlight the challenges for proteomic studies. The specific requirements and procedures for protein sample preparation of these different types of plant material are also described to help scientists overcome the unique challenges of working with plant samples.

7.2.1 Structural Proteome: From Plant Organ to Subcellular Compartment

Basically, plants have their organs splited in two categories: (i) vegetative (constituted of root, stem, and leaf) and (ii) reproductive (constituted of flower, fruit, and seed) [1]. Typically, the plant body consists of two organ systems, only.

- The root system (below the ground) which functions to anchor the plant to the soil, to absorb water and minerals, and to store some of the products of photosynthesis.

- The shoot system (above the ground) which includes the stem and conducts minerals and water from the root to the leaves with leaves being the critical location of photosynthesis for energy production and synthesis of organic and other compounds. At the reproductive stage, the shoot also includes flowers which when fertilized develop into fruit carrying the seeds for dispersion.

Because of their important role in energy production and reproduction, as well as because of their ready availability and ease of harvesting, the shoot system and its specific components are responsible for most proteomic publications. In this way, various works report the importance of some proteins for the correct functioning of the biological plant system [1]. An example is the ribulose-1,5-bisphosphate carboxylase/oxygenase commonly known by the abbreviation as RuBisCO. The RuBisCO is one of the enzymes involved in the Calvin cycle which is responsible for the photosynthesis process. The RuBisCO is involved directly in the carbon fixation, a process by which atmospheric carbon dioxide is converted by plants and other photosynthetic organisms to energy-rich molecules such as glucose. Another protein involved also in the energy metabolism of the plants are the heat shock proteins and ATP synthase alpha chain. These proteins act together in the adenosine triphosphate (ATP) synthesis and proton transport. The ATP synthesis is utmost important since it provides the necessary energy to the cell through the adenosine diphosphate (ADP)$^+$ inorganic phosphorus; both are joined together by the ATP synthase.

However, although the shoot system represents the main sample for understanding many biological processes of the plant, such system presents many analytical challenge because of the presence of high levels of phenolic compounds and derived pigments, and, therefore, it requires a good cleanup process or sample preparation. Some phenolic compounds can be widespread throughout the plant, while others are specific to certain plants, plant organs, and developmental stages and can be a major challenge for protein extraction.

The root system, in turn, is a simple structure that has been largely ignored in plant proteomics, having lost out to the shoot system. Only over the past years has interest in roots increased, because of their direct involvement in the perception of stresses related to water and nutrient availability in soils. Roots do not represent any greater difficulty in terms of protein extraction, though yields are somewhat lower than from leaves. Rather, issues arise with the isolation of clean, soil or media-free, roots [1].

Subcellular proteomics also is essential not only for understanding different roles of protein species and functions of subcellular compartments as a whole but also for unraveling the mechanism of protein targeting, trafficking, regulation, and dynamic changes that occur in plant processes. Proteomic studies have been conducted in several subcellular compartments of plant cells, including chloroplasts, mitochondria, nuclei, vacuoles, plasma membrane, and cell wall. The chloroplast and mitochondria are the two most thoroughly studied subcellular proteomes. In both cases, the organelles have been further fractionated into subproteomes, allowing detailed localization of proteins to different suborganelle compartments [16]. These analyses have helped to improve genome annotation, e.g., correcting errors in protein N- and C-terminus determination and intron/exon prediction.

7.2.2   Proteins Interaction and Their Functions for Plant Defense

Most proteins exert their functions through transient or stable interactions with other proteins. Such interactions play crucial roles in signaling, enzyme activity regulation, and plant defense [16]. For example, changes in proteins from leaf blades of rice plants infected with the blast fungus *Magnaporthe grisea* were recently investigated. Forty-five proteins out of 63 were identified. Some of these proteins were fairly homologous to those from rice plants, such as photosystem II oxygen-evolving complex protein 1, photosystem II oxygen-evolving complex protein 2, RuBisCO LSU, and RuBisCO SSU [38]. The photosystem II is the first protein complex in the light-dependent reactions of oxygenic photosynthesis. It is located in the thylakoid membrane of plants. Within the photosystem, enzymes capture photons of light to energize electrons that are then transferred through a variety of coenzymes and cofactors to reduce plastoquinone to platoquinol. The energized electrons are replaced by oxidizing water to form hydrogen ions and molecular oxygen. By replenishing lost electrons with electrons from the splitting of water, photosystem II provides the electrons for all of photosynthesis to occur. The hydrogen ions (protons) generated by the oxidation of water help to create a proton gradient that is used by ATP synthase to generate ATP. The energized electrons transferred to plastoquinone are used to reduce nicotinamide adenine dinucleotide phosphate ($NADP^+$) to NADPH or are used in cyclic photophosphorylation (Fig. 7.1) [42].

A targeted proteomics approach was used to identify the most abundant proteins from tomato xylem sap upon infection with *Fusarium oxysporum* [60]. This not only confirmed the presence of known pathogenesis-related (PR) proteins in the tomato sap but also the identification of a new PR-5 isoform. The PR proteins are involved in response to pathogens. mRNA level of the PR-5 gene is significantly changed after cutting the inflorescence stem indicating the existence of a network of signal transducing pathways as other stress-regulated genes [41].

Besides the analyses of the whole plant-pathogen proteome, several reports have also focused on elicitation/perception signaling mechanisms. For example, in the field of plant perception signaling, protein kinases play a central role during pathogen recognition and subsequent activation of plant defense mechanisms. In a study realized by Rakwal and Komatsu [59], it was observed the influence of exogenous jasmonic acid (JA) in defense mechanisms of rice (*Oryza sativa* L.). In the study, the proteins appearing or modified were identified through N-terminal and/or internal amino acid sequencing. Proteomics followed by immunological studies indicated that JA affects defense-related gene expression in rice seedlings, as evidenced by synthesis of novel proteins with potential roles in plant defense.

Other proteins have been identified in different compartments of different plants acting significantly in the defense process. For example, in the sunflower seed,

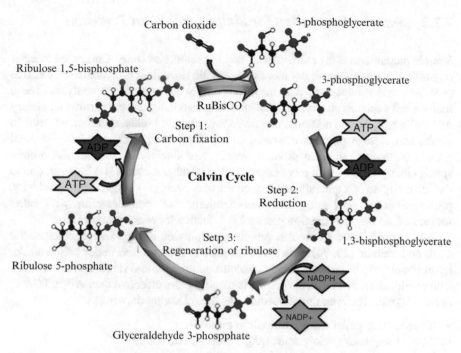

Carbon dioxide   3-phosphoglycerate

Ribulose 1,5-bisphosphate

                    RuBisCO   3-phosphoglycerate

ADP   ATP

Step 1:
Carbon fixation

ATP   **Calvin Cycle**   ADP

Step 2:
Reduction

Ribulose 5-phosphate   Setp 3:   1,3-bisphosphoglycerate
                    Regeneration of ribulose

                    NADPH

Glyceraldehyde 3-phospphate   NADP+

Fig. 7.1 Calvin cycle involving in the photosynthesis process, showing steps of carbon fixation, reduction, and regeneration of ribulose

putative dehydrin and aspartic proteinase were identified. Putative dehydrin is involved in response mechanisms in some abiotic stress such as dehydration and low temperatures. Regarding aspartic proteinase, it is involved in lipid metabolism processes [4]. In addition, 10 kDa late embryogenesis abundant protein was also found in some vegetables [6]. This protein is involved in the growth/cell division processes and is accumulated in the last stages of embryogenesis in seeds, related to processes of response to osmotic and water stresses. Another important protein for the correct functioning of the plant biological system is globulin seed storage protein (11 S helianthinin). This protein is considered the major group of storage proteins, being reported to account for 60% of the total proteins presented in the mature seed. Cytosolic phosphoglycerate kinase is involved in metabolic/energy processes, specifically in the carbohydrate metabolism. The 17.7 kDa heat shock protein has also been found in plant study; it is an important protein involved in stress responses, hyperosmotic pressures, heat, high light intensity, and hydrogen peroxide. Furthermore, this protein participates in the process of protein folding. Glutathione peroxidase and Mn-superoxide dismutase were also identified in transgenic soybean seeds [5]. These enzymes are involved in mechanisms of the oxidative stress response.

7.2.3   Sample Preparation for Simplifying Plant Proteome

Sample preparation is the main critical step in proteomics study. Compared to other organisms, plants are generally more problematic since their cells are usually rich in proteases and compounds that may interfere with proteomics analysis. These include cell wall polyphenols, polysaccharides, starch, lipids, and various secondary metabolites [4, 16]. In addition, certain tissues contain highly abundant proteins in leaves and storage proteins in seeds, which dominate protein profiles. For total protein extraction, an ideal protocol would reproducibly capture all the protein species in a proteome with low contamination of other molecules. However, due to the diversity of protein abundance, molecular weight, charge, hydrophobicity, posttranslational processing and modifications, and complexation with other molecules, no single extraction protocol is effective for every protein [16].

A commonly used protocol is protein precipitation using TCA (trichloroacetic acid) and acetone [20, 76]. It is based on protein denaturation under acidic and/or hydrophobic conditions that help to concentrate proteins and remove contaminants. The combination of TCA and acetone is usually more effective than either TCA or acetone alone. However, this procedure has the following drawbacks:

- The resulting pellet can be difficult to dissolve.
- TCA precipitates nucleic acids longer than 20 nucleotides.
- In some cases, TCA can hydrolyze proteins.

Another common protocol for plant material is phenol extraction that involves protein solubilization in the phenol phase, followed by precipitation with methanol and ammonium acetate [66]. This method has been shown to generate high-quality protein extracts from a variety of plant species. This feature meant it could be adapted for protein extraction from plant tissues [66], taking advantage of its ability to remove the interfering compounds, but also for its ability to remove DNA.

In this phenol extraction method, the first step consists of separating proteins from salts, nucleic acids, and carbohydrates with a Tris-phenol buffered solution mixed with a sucrose buffer. This fact inverts the phases putting the phenol phase at the top. The proteins, denatured and dissolved in the phenol phase, are then precipitated with a solution of methanolic ammonium acetate. The pellet is then successively washed with methanolic ammonium acetate, followed by acetone and methanol washes to remove pigments and lipids. However, this method can be more time-consuming, and the pellet can be difficult to resolubilize. In addition, it is important to note that:

- The phenol extraction method is more favorable for solubilizing membrane-associated proteins, not like the TCA/acetone method, which favors water-soluble proteins.
- For both the TCA/acetone and phenol extraction protocols, the use of protease inhibitors is not mandatory, because these methods use denaturing conditions, which have been reported to inhibit protease activity.

- Although manual grinding for tissue disruption leads in our hands to better protein yields, this step can be automated for higher throughput if laboratories are equipped with a bead beating tool which uses grinding balls and a shaking homogenizer to disrupt the tissues or cells. The use of these tools also provides better reproducibility between samples.
- Protein solubilization from the pellet is more efficient after phenol extraction than using the TCA/acetone method, resulting in reduced protein loss.

In order to enhance the coverage and detection of certain groups of proteins such as membrane proteins and low-abundant proteins, various strategies have been developed over the years to fractionate proteins into subproteomes based on biochemical, biophysical, and cellular properties, for example, (1) sequential extraction with a series of reagents based on differential protein solubility [49]; (2) phase partitioning using organic solvents or detergent Triton X-114 [56]; (3) liquid chromatography (LC) fractionation [10]; and (4) isolation of highly enriched organelles or subcellular compartments including chloroplasts, mitochondria, nuclei, vacuoles, peroxisomes, microsomal membranes, plasma membranes, cell walls, and the apoplast [16]. These frontend fractionation procedures have greatly improved detection and resolution by reducing the overall sample complexity and thus increase proteome coverage, e.g., the detection of low-abundant proteins.

Although RuBisCO plays a crucial role in the plant functioning, the presence of this protein for some study can be a challenge because its presence can make low proteins. Thus, some RuBisCO depletion strategies have been developed. Among them, small molecules such as phytate or polyethylenimine (PEI), which interact with specific proteins, have been used to precipitate RuBisCO from protein samples. In the case of phytate, the interaction, done in the presence of Ca^{2+} at a defined pH of 6.8, removed 85% of RuBisCO from soybean leaves [40]. PEI was successfully used in combination with fractionation to increase the protein resolution, an approach called PARC (PEI-assisted RuBisCO cleanup) [81]. However, these interactions are not specific to RuBisCO and can also precipitate non-targeted proteins.

In addition, Commercial kits using immunoaffinity removal of RuBisCO are also available (IgY RuBisCO columns from Sigma; RuBisCO depletion kit from Agrisera). However, the antibodies in such kits do not work well for every plant species. A different approach, using differential PEG (polyethylene glycol) precipitation, was used to isolate RuBisCO from a specific fraction [79]. PEG present in the fractions now containing only low levels of RuBisCO could then be cleaned up using one of the protein extraction methods used for plants.

7.2.4 Analytical Technique Applied for Uncover the Plant Proteome

For several years, SDS-PAGE has been one of the most widely used tools for the separation of total protein extracts as well as protein fractions obtained from various prefractionation procedures. However, for high-resolution protein profiling in biological systems, two-dimensional electrophoresis (2-DE), comprised of a first isoelectric focusing (IEF) step followed by sodium dodecyl sulfate polyacrylamide gel electrophoresis (SDS-PAGE) has been the selected method ever since it was developed in the early 70's and will continue to be used as a high-resolution protein separation technology [28].

As a powerful tool for surveying the proteome, 2-D gel technology offers high-resolution visualization of protein isoforms, degradation products, and potential PTMs and allows empirical estimation of protein molecular weight, pIs, and expression levels. Proteins separated by 2-D gel fit as good materials for fast protein identification and automated database searching. However, 2-D gel technology itself has limitations, e.g., relatively low throughput, no automation possible, and difficulty in resolving membrane proteins and very acidic or basic proteins. It is widely purported to be unsuitable for the separation of hydrophobic proteins because they often aggregate during IEF and do not readily enter the second-dimension SDS-PAGE [64]. Thus, LC is a technique alternative to 2-DE for protein separation, although both can be combined in a single experiment, when dealing with low-abundant proteins. LC has the advantage of using liquid phase compatible with different infusion into a mass spectrometer (LC-MS), although, unfortunately, proteins are not generically amenable to high resolution by reversed-phase chromatography (RP-HPLC) [11, 16].

Often referred to as "shotgun proteomics," methodologies using LC have the potential for correcting the sampling error that otherwise arises for proteins with unusually low or high molecular weight, low abundance, and/or high hydrophobicity. It is an automated high-throughput protein identification technology and has already been used effectively to catalog many polypeptides in total protein extracts from several organisms. In rice, over 2300 proteins were identified from a variety of tissues using multidimensional protein identification technology (MUDPIT) [16]. This method relies on sequence databases for assigning protein identities, and it works best with proteins from organisms for which complete genome sequence data are available. This approach generally provides no information about the original proteins, such as molecular weight, pI, modifications, and quantification. The quantitative limitation can be overcome by isotope labeling techniques, such as isotope-coded affinity tag (ICAT) or isobaric tags for relative and absolute quantification (iTRAQ), which provide good alternatives for quantitative proteomics.

The identification of multiprotein complex components and the analysis of protein-protein interaction are essential to understand most cellular processes. Protein interaction studies are carried out by using two-hybrid systems, developed by Fields and Song [23], protein chips, and the large-scale approach of tandem affinity purification (TAP)-MS. This last strategy has been used in the analysis of the yeast proteome, allowing the purification of more than 589 multiprotein assemblies [26], and can be applied to higher eukaryotes, avoiding the problem of the competition from corresponding endogenous proteins. Another method involves the use of fluorescence resonance energy transfer (FRET) between fluorescent tags on interacting proteins, by using green, cyan, and yellow fluorescent protein [57]. The great advantage of the approach is its application to in vivo analysis by microscopy.

In addition to protein identification, matrix-assisted laser desorption/ionization coupled mass spectrometry (MALDI-MS), and electrospray ionization mass spectrometry (ESI-MS) are useful techniques for the analysis of the accurate protein molecular weight, protein quantification, and PTM. Another technique that has been also used by many is Fourier transform ion cyclotron resonance mass spectrometry (FT-ICR-MS) because of its ability to provide the highest mass resolution (e.g., 500,000), the best mass accuracy (sub-ppm), and extremely high sensitivity of any MS technique. It allows the insertion of proteins present in a mixture direct to an ESI equipment and fragmented by a variety of gas-phase dissociation mechanisms.

7.3 Sample Preparation Focusing on Plant Metallomics

7.3.1 Defining Metallomics

Similar to genes/proteins, genomes/proteomes, and genomics/proteomics, and any other trilogy applied to biomolecules, one can define the relationship among metals, metallomes, and metallomics. In fact, metallomics area was coined recently by Prof. Hiroki Haraguchi, now professor emeritus, Nagoya University, and it is defined as the science of biometals [30], and it congregates multidisciplinary concepts [67]. In fact, the metallomics area involves not only the identification/quantification of the metal/metalloid present in a cell or tissue type but also the identification of what form it is present, the biomolecule to which they are bound, and the coordination groups involved [52]. As noted, metallomics can provide new opportunities for discoveries.

7.3.2 Sample Preparation for Plant Metallomics

In terms of sample preparation for plant metallomics, the first tough, which should come in mind when the focus is the sample preparation, is the compromise between the efficiency of such process and the maintenance of the metal-protein bound, once metal-binding sites, metal-dependent structural or conformational changes, and metal stoichiometry are currently the main targets of the study [52].

This statement reinforces the needed care in using those classical separation techniques, currently applied in proteomics studies, such as SDS-PAGE or 2-D-PAGE. Although attractive, such techniques currently require reducing agents (as β-mercaptoethanol or dithiothreitol (DTT)), and the SDS-PAGE causes protein denaturation, resulting in the aggregation of individual polypeptides, as well as in the disruption of the protein-bound complexes [46]. Then it is easy to rationalize that such procedures come in the opposite way to those gentle ones, which are utmost necessary to maintain the integrity of the metal-protein bound. In this way, when the metallomics information has to extract directly from a target protein present in a gel spot, then nondenaturing strategies are preferable [7, 18, 65]. In fact, the metal lost during the sample preparation step will be influenced by the affinity between the metal and a given protein. Then, when the interactions between metals and proteins present low stability constants, the metals will be easily lost, while at high values of the stability constants they will not. These differences in metal-protein stabilities may explain the success of several reports in the literature on the analysis of metalloproteins, even using denaturing conditions for protein separation [17, 55, 58]. In this way, as a general approach, two possibilities must be considered [52]. The first one is the extraction of the entire sample proteome (without caring for the metal-protein bonds) followed by the isolation of the set of proteins with affinity to a particular metal and their identification by one of the canonical proteomics approaches, and the second one is the extraction of intact metal-protein complexes followed by their separation and identification. The emphasis, in the latest strategy, is on the preservation of the metal-protein complex.

One of the major problems found in plant protein and enzyme extractions is the presence of phenolic compounds released during tissue grinding. These phenols are readily oxidized to quinones by plant enzymes, and the phenolic compounds, as quinones, either react with the proteins, inactivating the enzymes, or change the mobility of protein molecules. Then, substances that avoid both oxidation and complexation to phenols/quinones, such as β-mercaptoethanol, are strongly recommended to be added to ground samples as well as to the extracting buffer [47].

It is important to note that extraction protocols greatly influence the preservation of the metal-protein bond. In this way, Sussulini et al. [69] evaluated two protocols for protein extraction from soybean seeds, focusing on metalloprotein analysis. One of them was based on hexane to ground the soybean and the extraction of the proteins carried out with Tris-HCl and β-mercaptoethanol. The other was based on petroleum ether and the extraction of the proteins with Tris-HCl, DTT, PMSF, SDS, and KCl. Since the petroleum ether presents higher numbers of carbon in its structure than hexane, its contribution to the extraction is higher, and the buffer containing KCl minimizes the counterion effects, improving the solubility of the proteins, and maintains the ionic strength of the medium. Then, the second protocol was chosen, and the metals/metalloids bound to soybean seed proteins contained in electrophoresis gels were mapped using synchrotron radiation x-ray fluorescence (SR-XRF) and determined by inductively coupled plasma optical emission spectrometry (ICP OES) and ICP-MS.

Besides the extraction medium, the use of different techniques for solubilizing the (metallo)proteins, such as microwave or ultrasound, is currently made. In this way, Magalhães and Arruda [47] evaluated 11 different extraction procedures, some of them using ultrasound extraction. Although it is currently used in many proteomic protocols for sample solubilization [46], the ultrasound energy may present some drawbacks when the focus is the metalloprotein solubilization. In fact, such strategy presented the poorest result in terms of the preservation of the integrity between metal and protein.

In another example involving sample preparation to plant metallomics, the use of microwaves as an alternative energy was considered by Sussulini et al. [70] for evaluating soybean through ionomic approach. In fact, the microwave technology was used for the total decomposition of the protein spots present in the 2-D gels. Additionally, in order to optimize the method for low sample mass amount, once the spot presents only few micrograms of mass, an elegant alternative was then proposed, which was based on the minivials concept [9, 50]. In this way, the protein spots were cut from the gel and dried at 40 °C at constant mass. The spots were then weighed (0.4–3 mg) using an autobalance and placed in a 1.8-ml cryovial polypropylene minivial. After a pre-reaction with 200 µl of concentrated nitric acid + 150 µl of 30% (v/v) H_2O_2, the minivials were capped and placed in a PTFE holder (capacity of four minivials per holder). The holder was then placed into the PTFE microwave vessel containing 15 ml of deionized water to keep the equilibrium pressure. In this way, the entire set was heated for 30 min in a microwave oven. At the end of the decomposition process, the volume of the minivials was completed to 1 ml with deionized water, and the analytes and residual carbon content (RCC) determined. Accurate determinations were found for calcium, copper, and iron from cytochrome P450, photosystem Q(B), and cytochrome c subunit.

The same strategy was applied for a comparative metallomics study involving transgenic and nontransgenic soybean seeds [71], which corroborated that not only the proteome was changed by the genetic modification but also the metallome. The amount of Ca^{2+}, Cu^{2+}, and Fe^{2+} was then determined in proteins ranging from 13.98 to 54.87 kDa, and significative differences were found to Fe^{2+} when transgenic and nontransgenic soybeans were considered.

Another strategy for attaining ionomics approach is the use of a laser ablation (LA) chamber coupled to the ICP-MS (LA-ICP-MS). For such task, PAGE techniques are currently used, and due to the denaturing nature of such techniques, native PAGE (or nondenaturing PAGE) has been explored for circumventing this problem, since the tertiary structure of the proteins is preserved during separation. However, this may not be true at all. For example, Jiménez et al. [32] reported some pitfalls due to metal losses in PAGE separation prior to LA-ICP-MS analysis. The authors studied two different metalloproteins: thyroglobulin, containing I, and superoxide dismutase (SOD), containing Cu and Zn. Quantitative detection of iodine in thyroglobulin using either SDS-glycine PAGE or native PAGE followed by LA-ICP-MS analysis was achieved. However, detection of Cu^+ and Zn^+ in SOD by LA-ICP-MS depended on the conditions of the PAGE method, since they were only detected (but not quantitatively) using SDS-tricine as the PAGE separation buffer. Cu^+ and Zn^+ losses were observed if glycine was used as the trailing ion in the PAGE separation, owing to the formation of Cu^+ and Zn^+ complexes. In both cases, native PAGE generated similar results to those achieved by the SDS-glycine method.

The plant ionomics approach involving LA-ICP-MS, but in the imaging mode, was developed by [45], for evaluating some differential protein species in transgenic (T) and nontransgenic (NT) *Arabidopsis thaliana* plants after their cultivation in the presence or absence of sodium selenite. For this task, a simple sample preparation procedure was performed, once the leaves were dried until constant mass at 38 °C on a stove. Then, the samples were selected randomly, and the distribution of Se and S is evaluated by laser ablation (imaging) ICP-MS, which consists in a quadrupole-based ICP-MS coupled with a laser ablation system. After the drying process, the leaves were fixed onto acetate double-sided adhesive tape and placed into the ablation chamber for analysis. The $^{13}C^+$ was used as internal standard for avoiding any fluctuation of the plasma source and the ablation process, and for circumventing polyatomic interferences in the ICP-MS measurements, the Se was evaluated as SeO^+ (at m/z 96) and S as SO^+ (at m/z 48), which was formed in the DRC cell. Through the results, it was possible to corroborate the detoxication mechanism for Se elimination in plants, once that great amount of this element was achieved in such plant compartment. Additionally, differential proteins found, such as glyceraldehyde-3-phosphate dehydrogenase, also present cysteine residues, which may have been substituted by selenocysteine.

Plant ionomics was also explored in the work of [43], by evaluating sunflower growth under different cultivation conditions after adding ca. 50, 350, and 700 mg of cadmium (as cadmium chloride) during the cultivation period. Treated and non-treated plants were compared in terms of some monitored ions ($^{63}Cu^+$, $^{56}Fe^+$, $^{31}P16O^+$, $^{24}Mg^+$, $^{55}Mn^+$, $^{64}Zn^+$) using ICP-MS. Microwave technology, through an optimized program as 5 min at 400 W, 8 min at 790 W, 4 min at 320 W, and 2 min at 0 W was used for sample preparation. Then, the decomposed samples were cooled and diluted to 25 mL with 1% (v/v) HNO_3 and the total content of elements determined by ICP-MS. This study revealed that, besides hyperaccumulator, sunflowers are Cd-hypertolerant plant species, growing in soils contaminated with high levels of cadmium. Contrary to other Cd hyperaccumulators, sunflower has the ability to accumulate high concentrations of cadmium in shoots without compromising its development.

A metalloproteomics approach was carried out by Chacón-Madrid et al. [12], reporting the evaluation of the *Arabidopsis thaliana*, genetically modified or not, which were cultivated in the absence or in the presence of different concentrations of sodium selenite. The metalloproteins were identified through LC-ICP-MS and ESI-MS/MS. For such task, the leaves of the *A. thaliana* were ground manually in a mortar with liquid N_2 and dried to constant mass at 37 °C. For total selenium determination through ICP-MS, the microwave oven technology was used (5 min at 400 W, 8 min at 790 W, 4 min at 320 W, 3 min at 0 W), using nitric acid (6 mL) and hydrogen peroxide (0.5 mL). Then, the samples were gently heated for the elimination of the excess of HNO_3.

In the protein extraction from *A. thaliana* leaves focusing on metalloproteomics approach, 1.00 g (fresh weight) of leaves was ground with liquid N_2 and treated with 5.0 mL of 30 mmol L^{-1} Tris-HCl (pH 7.5). The mixture was sonicated for 15 min and subsequently centrifuged at 8000 g for 10 min at 4 °C. Proteins were quantified in the supernatants using a 2D Quant Kit®. Then, they were filtered through a 0.22 μm membrane and analyzed by LC-ICP-MS.

For obtaining information regarding possible proteins present from the chromatographic Se-containing peaks, fractions of leaf extracts of N-GM and GM were then collected, lyophilized, digested with trypsin, and analyzed through mass spectrometry (ESI-MS/MS). Identified protein species were classified according to their biological activities as energy, such as ribulose bisphosphate carboxylase small chain 1A chloroplastic, ribulose bisphosphate carboxylase small chain 2B chloroplastic (in the fraction 1), and photosystem I reaction center subunit IV A chloroplastic (in the fraction 2).

Through the results, the Se accumulation in the GMO was higher than nongenetically modified leaves indicating that the genetic modification itself is a stressful event to the plants.

7.4   Sample Preparation Focusing on Plant Metabolomics

7.4.1   *Metabolomics and Metabolite Analysis Workflow*

Metabolomics is considered as a relatively new research field, which is focused on the measurement of all the metabolites, through qualitative and quantitative analysis, in the metabolome from a specific system/organism under certain conditions. In addition, metabolomics is the complement to other areas; among them are genomics, transcriptomics, and proteomics, offering a global and targeted assessment of the physiological state of a biological system [21, 53, 74]. Since metabolites play crucial roles in the living system, acting in cellular and physiological processes, it is of utmost importance to evaluate different strategies for metabolite extraction [29, 51].

Such evaluation is required for an accurate metabolite profiling study, considering the general system complexity. In fact, a solvent combination, which is suitable for one chemical class, cannot be appropriated for another one and also may not be suitable to extract a large number of metabolites. For plants and animals, the metabolite extraction from tissue samples may represent an even greater challenger, considering the variety of metabolites, the diversity of their physicochemical properties, and concentration ranges in solid samples [48, 51]. This explains why there is no ideal method/solvent for metabolite extraction to achieve global metabolic profile. In this way, a sort of tests involving different solvents followed by extraction method optimization is needed to solubilize a higher number of metabolites from plant tissues to access the global metabolic profile [21, 48].

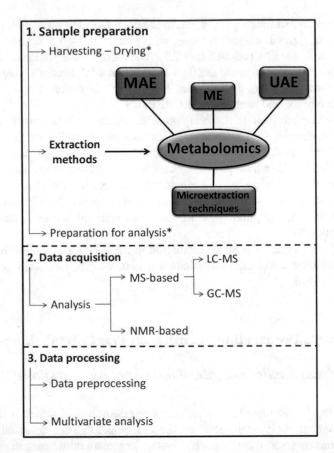

Fig. 7.2 Metabolomic experimental workflow focusing on sample preparation step with the most important extraction methods. (Adapted from Kim and Verporte (2010). *These steps are not necessary in a certain analysis)

Currently, metabolomic analysis is consisted of three different experimental parts (Fig. 7.2). Sample preparation is the first step, followed by data acquisition, and the final step is related to data mining using suitable chemometric approaches [48]. For data acquisition, analytical chemical techniques are employed, mainly liquid or gas chromatography coupled to mass spectrometry (MS) and nuclear magnetic resonance (NMR) [8]. Although all these three steps are strongly correlated, sample preparation is the critical step for the accuracy of the results, once the sample preparation approaches are dependent on the metabolite/solvent properties [35, 36].

The main purposes of plant metabolite extraction are isolation and identification of the natural compounds positively correlated to any human's health [24]. Distinct extraction methods have been used to extract metabolites from plant tissues; among them are maceration [54, 73], microextraction techniques [8, 80], microwave-assisted extraction [19, 72, 82], and ultrasound-assisted extraction [54, 61]. Considering the relevance, in terms of applicability and efficiency, the last two extraction methods will be emphasized and discussed from now.

7.4.2 Microwave-Assisted Extraction

Microwave-assisted extraction (MAE) is an efficient strategy that has emerged as a potential extraction technique for varied kinds of biological samples, due to distinct aspects. One of the most important advantages of MAE is the capability to extract valuable compounds in a shorter time employing low usage of extraction solvents [21, 72, 82]. The merits of MAE comprise a simple and rapid sample preparation, controllable temperature, reduction of time, and high extraction yields [22, 27]. In addition, MAE increases the dissolution rate of metabolites, due to an increase in the internal temperature, and such technique is quite adaptable on a small or large scale [21, 82]. However, an issue of such method is the incompatibility with some organic solvents because of the risk of explosion [21]. In this way, water is an interesting alternative for MAE [27], since the amount of water present in solvent, as concentration of aqueous solution, impacts significantly the extraction yield [31].

Currently, considering MAE advantages, such technique has been employed in different research areas, mainly in plant and food metabolomics [22, 27, 77, 78]. MAE of secondary metabolites of plants may be affected by several factors, such as power and frequency of microwave, time of microwave radiation, extraction pressure and temperature, number of extraction cycles, type and concentration of solvent, ratio of solid to liquid, and particle size of plant samples, among others [82]. One of the most important parameters for MAE is its association to the solvent composition, since it affects the solubility of the target compounds and microwave energy absorption, which is determined by solvent dissipation factor. The solvent choice selection is currently carried out not only considering its affinity with the target component but also its capability in absorbing microwave energy [15, 82].

Fang et al. [22] developed a simple and rapid method using MAE combined with HPLC-DAD-ESI-MS/MS for the simultaneous extraction, identification, and quantification of phenolic compounds in a common vegetable (*Eclipta prostrata*) in China. Several relevant parameters related to MAE were evaluated, among them type of solvent, microwave power, temperature, ratio of solid to liquid, and extraction time. Six phenolic acids, six flavonoid glycosides, and one coumarin were firstly identified, as well as phenolic compounds quantified by HPLC-DAD with linearity, precision, and accuracy. The most suitable solvent for MAE was achieved by comparing distinct mixed ratio of methanol/water and ethanol/water solvent systems. The highest yields were obtained employing 50% (v/v) ethanol as solvent. In addition, the authors also evaluated the ultrasound-assisted extraction (UAE) and heat reflux extraction (HRE) for the extraction of the phenolic compounds. According to the results, the extract obtained by MAE showed the strongest antioxidant activity, and it avoids the degradation of dicaffeoylquinic acids and wedelolactone, obtaining the highest yields of all components in a shorter amount of time and using less ethanol and energy [22].

A comparison between MAE and maceration (ME) systems for the extraction of valuable compounds from walnut (*Juglans regia* L.) leaves was carried out by Vieira et al. [73]. For this task, an experimental design, assisted by response surface methodology, was developed to optimize important variables of the ME and MAE. The criteria considered for defining the optimized conditions were the quantification of some metabolites by HPLC-DAD, the extraction yield, and the spectrophotometric results of total phenolics and flavonols. In general, high extraction yields were obtained for MAE process, except for 3-O-caffeoylquinic acid, since ME system was more efficient to extract such phenolic acid. After optimizing all the MAE conditions (3.0 min of extraction time, temperature of 107.5 °C, and 67.9% of ethanol concentration), significant reductions in the extraction time followed by high metabolite extraction yields were observed for MAE in comparison with ME system [73].

MAE in combination with HPLC-UV was proposed by Wei et al. (2015) for simultaneous quantification of six bioactive alkaloids in *Menispermum dauricum* DC plants. After standard compound isolation, the optimal MAE condition was obtained at 60 °C with 70% (v/v) of ethanol concentration as the extracting solvent, and the solvent to solid ratio was 20:1. Chromatographic separation was carried out on a reversed-phase C_{18} column (250 × 4.6 mm, i.d., 5 μm) in a gradient elution with mobile phase consisting of A (1% of aqueous formic acid) and B (acetonitrile containing 1% of formic acid) at a flow rate of 1.5 mL min^{-1}. At the end of the optimization and validation, MAE-HPLC-UV method required 11 min to extract and 18 min to quantify six different alkaloids in different plant compartments, achieving quantitative extraction recoveries with acceptable sensitivity and precision (Wei et al. 2015).

In the same way, Koyu et al. [39] investigated MAE as a potential technique for metabolite extraction focused on bioactive constituents in mulberry (*Morus nigra* L.) fruits. In addition to the optimization of MAE, such strategy was compared to conventional methods, in this case ultrasonic bath and orbital shaker. The MAE optimization was performed through a surface response of MAE parameters, such as microwave power, ethanol concentration, and time of extraction. The optimization was performed in order to maximize bioactive metabolites and tyrosinase inhibitory activity. Ethanol concentration and extraction time were considered as the most influential parameters regarding extraction of total phenol content, while ethanol concentration was the most efficient in regard to tyrosinase inhibitory activity. After such step, total phenol, flavonoid, and anthocyanin contents were quantified by spectrometry and UHPLC-DAD-ESI-MS/MS. According to the results, tyrosinase inhibitory activity was better achieved for MAE extracts in the optimized condition (500 W of microwave power, 35% of ethanol concentration, 10 min of extraction time) when compared to conventional methods. Additionally, MAE process significantly decreased the extraction time, which is welcome for bioactive contents analysis, since plant extracts contain various metabolites both active and non-active.

Thus, MAE technique may be pointed out as a more advanced technique by providing greater bioactive content and bioactivity with reducing demand for organic solvent, energy, and time [39].

7.4.3 Ultrasound-Assisted Extraction

Ultrasound-assisted extraction (UAE) has emerged as an efficient extraction technique which demands simple manipulation and results in high yield of analyzed compounds in short times. In addition, in this method, reduced volumes of solvent and lower energy are needed, meeting the requirements of "green chemistry" [14]. The cavitation effect enables to enhance the extraction efficiency of organic compounds by ultrasound, and such effect may cause locally high temperatures and pressures, resulting in a quickly isolation of extracted compounds [34, 62, 68]. The efficient release of extracted compounds from the matrix into the extraction medium is achieved by the ultrasound penetration in the matrix, rupturing the cell walls [54, 75].

Ultrasonication enables the swelling and hydration of plant materials, consequently enlarging the pores of the cell wall. Hence, the rate of mass transfer and break the cell walls are improved, increasing the extraction efficiency. In addition, UAE is very simple and rapid and has no limitations related to extraction solvent selection [36]. When compared to MAE, a broad range of solvents may be considered in UAE, such as methanol, ethanol, isopropyl alcohol, and their mixture [13]. Additionally, UAE can reduce 1–2 times the extraction time, saving energy and costs, and it may reduce metabolite degradation [61]. However, since high ultrasonic energy can result in high temperatures, such parameter must be firstly evaluated to avoid the degradation of metabolites [27, 37].

UAE combined with LC-ESI-MS/MS was considered by Oniszczuka and Olech [54] for the analysis of phenolic acids from kale (*Brassica oleracea* L. var. *sabellica*), since vegetables belonging to this group are an important source of bioactive metabolites. Distinct parameters regarding UAE were evaluated; among them are four solvents (ethanol, 80% of aqueous ethanol, methanol, and 80% aqueous methanol) and two extraction times (40 and 60 min), with ultrasound frequency of 20 kHz and power of 100 W. The optimized condition for effective phenolic acid isolation from kale was 80% aqueous ethanol and extraction time of 60 min. In the same way, as previously discussed, the results obtained by Oniszczuka and Olech [54] indicate that each factor in UAE has a significant influence on the yield of metabolite extraction from plants. In this work, the solvent type and extraction time influenced extraction yield under ultrasound treatment. Concerning LC-ESI-MS/MS analysis, several metabolites were identified and quantified in the kale extracts through a simple manipulation, employing reduced volume of solvent, with high reproducibility, meeting the requirements of "green chemistry" [54].

A fast, comprehensive, and reproducible one-step extraction method for the rapid analysis of polar and semipolar metabolites and other biomolecules in

Arabidopsis thaliana seeds was proposed by Salem et al. [63]. Two distinct extraction solvents were considered as follows: (1) methyl *tert*-butyl ether (MTBE) and methanol, 3:1 (v/v) (M1), and (2) water and methanol, 3:1 (v/v) (M2). For metabolomic evaluation, after M1 solvent extraction addition, the tubes were thoroughly vortexed for 1 min and, then, incubated on an orbital shaker (100 rpm) for 45 min at 4 °C. To improve biomolecule extraction, a sonication step for 15 min was also considered. Next, a liquid/liquid phase separation was achieved by M2 solvent adding followed by vortexing and centrifugation at a speed of 20,000 g for 5 min at 4 °C. According to the authors, this one-step extraction protocol allows the analysis of polar, semipolar, and hydrophobic metabolites, as well as insoluble or precipitated compounds, including proteins, being such method suitable for "multi-omics" sample extraction, preparation, and analysis. Regarding metabolomic evaluation, more than 100 primary metabolites and 50 secondary metabolites were measured employing such one-step extraction strategy [63].

Since the extraction solvent is an important aspect for metabolite extraction, Mahmud et al. [48] compared three different strategies employing a UAE step (15 min), for global metabolite extraction from soybean leaves using ultrahigh-performance liquid chromatography coupled with high-resolution mass spectrometry (UHPLC-HR-MS). The evaluated solvents were ammonium acetate/methanol (AAM), water/methanol (WM), and sodium phosphate/methanol (PM). According to the results, both methods using AAM and WM covered a wider range of metabolites, providing greater detection of molecular features than the method using PM. Through a multivariate statistical tool, several clustering analyzes showed both methods using AAM and WM as a tight and overlapping strategy when compared to method with PM. In addition, statistically significant score plot separation was observed between AAM vs. PM and WM vs. PM, as no significant separation was observed between AAM vs. WM. Additionally, a larger number of extracted metabolites were achieved at a significantly higher level using AAM vs. PM, suggesting ammonium acetate/methanol as more efficient and suitable solvents for LC-MS-based plant metabolomics profile study [48].

7.5   Conclusions

From the information presented in this chapter, it is clear the importance of a well-established sample preparation method for "omics" area, in order to attain reliable and significant results. Due to the complexity of the analytes (proteins, enzymes, metabolites) in terms of their structures and instabilities, the use of gentle protocols for preserving their identities is of great importance along the sample preparation process. Then, it is necessary that sample preparation be currently rediscussed for attaining the necessities nowadays in terms of speed, accuracy, costs, and amount of the analytes, among others. This is extremely salutary to the progress of omics area, which depends intrinsically of a well-performed sample preparation.

Acknowledgments The authors gratefully acknowledge the financial support of Fundação de Amparo à Pesquisa do Estado de São Paulo (FAPESP), the Conselho Nacional de Desenvolvimento Científico e Tecnológico (CNPq), and Financiadora de Estudos e Projetos (FINEP).

References

1. Alvarez S, Naldrett MJ (2016) Plant structure and specificity - challenges and sample preparation consideration for proteomics. In: Mirzaei H, Carrasco M (Eds.) Modern proteomics – Sample preparation, analysis and practical applications. Advances in Experimental Medicine and Biology, vol 919. Springer, Cham, pp 63–82
2. Arruda MAZ (2007) Trends in sample preparation, 1st edn. Nova Science, New York, p 304
3. Baginsky S (2009) Plant proteomic: concept, application, and novel strategies for data interpretation. Mass Spectrom Rev 28:93–120
4. Barbosa HS, Souza DLQ, Koolen HHF, Gozzo FC, Arruda MAZ (2013) Sample preparation focusing on plant proteomics: extraction, evaluation and identification of proteins from sunflower seeds. Anal Methods 5(1):116–123
5. Barbosa HS, Arruda SCC, Azevedo RA, Arruda MAZ (2012) New insights on proteomics of transgenic soybean seeds: evaluation of differential expressions of enzymes and proteins. Anal Bioanal Chem 402(1):299–314
6. Battaglia M, Covarrubias AA (2013) Late Embryogenesis Abundant (LEA) protein in legumes. Front Plant Sci 4(6):1–11
7. Berrada W, Naya A, Iddar A, Bourhim N (2002) Purification and characterization of cytosolic glycerol-3-phosphate dehydrogenase from skeletal muscle of jerboa (*Jaculus orientalis*). Mol Cell Biochem 231:117–127
8. Bojko B, Reyes-Garcés N, Bessonneau V, Goryński K, Mousavi F, Silva EAS, Pawliszyn J (2014) Solid-phase microextraction in metabolomics. Trends Anal Chem 61:168–180
9. Brancalion ML, Arruda MAZ (2005) Evaluation of medicinal plant decomposition efficiency using microwave ovens and mini-vials for Cd determination by TS-FF-AAS. Michrochimica Acta 150:283–290
10. Brown JWS, Flavell RB (1981) Fractionation of wheat gliadin and glutenin subunits by two-dimensional electrophoresis and the role of group 6 and group 2 chromosomes in gliadin synthesis. Theor Appl Genet 59(6):349–359
11. Cánovas FM, Dumas-Gaoudot E, Recorbet G, Jorrin J, Mock HP, Rossignol M (2004) Plant proteome analysis. Proteomics 4(2):285–298
12. Chacón-Madrid K, Pessôa GS, Salazar MM, Pereira GAG, Carneiro GMT, Lima TB, Gozzo FC, Arruda MAZ (2017) Evaluation of genetically modified *Arabidopsis thaliana* through metallomic and enzymatic approaches focusing on mass spectrometry-based platforms. Int J Mass Spectrom 418:6–14
13. Chemat F, Rombaut N, Sicaire AG, Meullemiestre A, Fabiano-Tixier AS, Abert-Vian M (2017) Ultrasound assisted extraction of food and natural products. Mechanisms, techniques, combinations, protocols and applications. A review. Ultrason Sonochem 34:540–560
14. Chemat F, Tomao V, Virot M (2008) Ultrasound-assisted extraction in food analysis. In: Ötles S (ed) Handbook of food analysis instruments. CRC Press, Boca Raton, pp 85–99
15. Chen L, Jin H, Ding L, Zhang H, Li J, Qu C, Zhang H (2008) Dynamic microwave-assisted extraction of flavonoids from Herba Epimedii. Sep Purif Technol 59(1):50–57
16. Chen S, Harmon AC (2006) Advances in plant proteomics. Proteomics 6(20):5504–5516
17. Chen SX, Shao ZZ (2009) Isolation and diversity analysis of arsenite-resistant bacteria in communities enriched from deep-sea sediments of the Southwest Indian Ocean Ridge. Extremophiles 13:39–48

18. Chevreux S, Roudeau S, Fraysse A, Carmona A, Devès G, Solari PL, Mounicou S, Lobinski R, Ortega R (2009) Multimodal analysis of metals in copper-zinc superoxide dismutase isoforms separated on electrophoresis gels. Biochimie 91:1324–1327
19. Dahmoune F, Nayak B, Moussi K, Reminia H, Madani K (2015) Optimization of microwave-assisted extraction of polyphenols from *Myrtus communis* L. leaves. Food Chem 166:585–595
20. Damerval C, Vienne D, Zivy M, Thiellemnt H (1986) Technical improvements in two-dimensional electrophoresis increase the level of genetic variation detected in wheat-seedling proteins. Electrophoresis 7(1):52–54
21. Ernst M, Silva DB, Silva RR, Vêncio RZN, Lopes NP (2014) Mass spectrometry in plant metabolomics strategies: from analytical platforms to data acquisition and processing. Nat Prod Rep 31:784–806
22. Fang X, Wanga J, Hao J, Li X, Guo N (2015) Simultaneous extraction, identification and quantification of phenolic compounds in *Eclipta prostrata* using microwave-assisted extraction combined with HPLC–DAD–ESI–MS/MS. Food Chem 188:527–536
23. Fields S, Song O (1989) A novel genetic system to detect protein-protein interactions. [Yeast two hybrid]. Nature 340(6230):245–246
24. Fiol M, Adermann S, Neugart S, Rohn S, Mügge C, Schreiner M, Krumbein A, Kroh LW (2012) Highly glycosylated and acylated flavonols isolated from kale (*Brassica oleracea* var. *sabellica*)–structure–antioxidant activity relationship. Food Res Int 47(1):80–89
25. Flis P, Ouerdane L, Grillet L, Curie C, Mari S, Lobinski R (2016) Inventory of metal complexes circulating in plant fluids: a reliable method based on HPLC coupled with dual elemental and high-resolution molecular mass spectrometric detection. New Phytol 211:1129–1141
26. Gavin AC, Bosche M, Krause R, Grandi P, Marzioch M, Bauer A, Schultz J, Rick JM, Michon AM, Cruciat CM, Remor M, Hofert C, Schelder M, Brajenovic M, Ruffner H, Merino A, Klein K, Hudak M, Dickson D, Rudi T, Gnau V, Bauch A, Bastuck S, Huhse B, Leutwein C, Heurtier MA, Copley RR, Edelmann A, Querfurth E, Rybin V, Drewes G, Raida M, Bouwmeester T, Bork P, Seraphin B, Kuster B, Neubauer G, Superti-Furga G (2002) Functional organization of the yeast proteome by systematic analysis of protein complexes. Nature 415(6868):141–147
27. Gong Z-G, Hu J, Wu X, Xu Y-J (2017) The recent developments in sample preparation for mass spectrometry-based metabolomics. Crit Rev Anal Chem 47(4):325–331
28. Görg A, Weiss W, Dunn MJ (2004) Current two-dimensional electrophoresis technology for proteomics. Proteomics 4(12):3665–3685
29. Hao J, Liebeke M, Astle W, Iorio MD, Bundy JG, Ebbels TMD (2014) Bayesian deconvolution and quantification of metabolites in complex 1D NMR spectra using BATMAN. Nat Protoc 9(6):1416–1427
30. Haraguchi H (2004) Metallomics as integrated biometal science. J Anal At Spectrom 19:5–14
31. Hemwimon S, Pavasant P, Shotipruk A (2007) Microwave-assisted extraction of antioxidative anthraquinones from roots of *Morinda citrifolia*. Sep Purif Technol 54:44–50
32. Jiménez MS, Gomez MT, Rodriguez L, Martinez L, Castillo JR (2009) Some pitfallsin PAGE-LA-ICP-MS for quantitative elemental speciation of dissolved organic matter and metallomics. Anal Bioanal Chem 393:699–707
33. Job D, Haynes PA, Zivy M (2011) Plant proteomics. Proteomics 11(9):1157–1158
34. Khan MK, Abert-Vian M, Fabiano-Tixier AS, Dangles O, Chemat F (2010) Ultrasound-assisted extraction of polyphenols (flavanone glycosides) from orange (*Citrus sinensis* L.) peel. Food Chem 119:851–858
35. Kim HK, Choi YH, Verpoorte R (2010) NMR-based metabolomic analysis of plants. Nat Protoc 5(3):536–549
36. Kim HK, Verpoorte R (2010) Sample preparation for plant metabolomics. Phytochem Anal 21:4–13
37. Klein-Júnior LC, Viaene J, Salton J, Koetz M, Gasper AL, Henriques AT, Vander Heyden Y (2016) The use of chemometrics to study multifunctional indole alkaloids from *Psychotria nemorosa* (*Palicourea comb. nov.*). Part I: extraction and fractionation optimization based on metabolic profiling. J Chromatogr A 1463:60–70

38. Konishi H, Ishiguro K, Komatsu SA (2001) Proteomics approach towards understanding blast fungus infection of rice grown under different levels of nitrogen fertilization. Proteomics 1(9):1162–1171
39. Koyu H, Kazan A, Demir S, Haznedaroglu MZ, Yesil-Celiktas O (2018) Optimization of microwave assisted extraction of *Morus nigra* L. fruits maximizing tyrosinase inhibitory activity with isolation of bioactive constituents. Food Chem 248:183–191
40. Krishnan HB, Natarajan SS (2009) A rapid method for depletion of Rubisco from soybean (*Glycine max*) leaf for proteomic analysis of lower abundance proteins. Phytochemistry 70(17–18):1958–1964
41. Li S, Strid Å (2005) Anthocyanin accumulation and changes in CHS and PR-5 gene expression in *Arabidopsis thaliana* after removal of the inflorescence stem (decapitation). Plant Physiol Biochem 43(6):521–525
42. Loll B, Kern J, Saenger W, Zouni A, Biesiadka J (2005) Towards complete cofactor arrangement in the 3.0 Å resolution structure of photosystem II. Nature 438(7070):1040–1044
43. Lopes Júnior CA, Mazafera P, Arruda MAZ (2014) A comparative ionomic approach focusing on cadmium effects in sunflowers (*Helianthus annuus* L.). Environ Exp Bot 107:180–186
44. Lopes Júnior CA, Barbosa HS, Galazzi RM, Koolen HHF, Gozzo FC, Arruda MAZ (2015) Evaluation of proteome alterations induced by cadmium stress in sunflower (*Helianthus annuus* L.) cultures. Ecotoxicol Environ Saf 119:170–177
45. Maciel BCM, Barbosa HS, Pessôa GS, Salazar MM, Pereira GAG, Gonçalves DC, Ramos CHI, Arruda MAZ (2014) Comparative proteomics and metallomics studies in *Arabidopsis thaliana* leaf tissues: evaluation of the selenium addition in transgenic and nontransgenic plants using two-dimensional difference gel electrophoresis and laser ablation imaging. Proteomics 14:904–914
46. Arruda MAZ, Magalhães CS, Garcia JS, Lopes AS, Figueiredo EC (2007) Strategies for Sample Preparation Focusing Biomolecules Determination/Characterization. In: Arruda MAZ (Ed.) Trends in Sample Preparation, 1st ed. Nova Science Publishers, New York, pp 245–288
47. Magalhães CS, Arruda MAZ (2007) Sample preparation for metalloprotein analysis: a case study using horse chestnuts. Talanta 71:1958–1963
48. Mahmud I, Sternberg S, Williams M, Garrett TJ (2017) Comparison of global metabolite extraction strategies for soybeans using UHPLC-HRMS. Anal Bioanal Chem 409:6173–6180
49. Molloy MP, Herbert BR, Walsh BJ, Tyler MI, Traini M, Sanchez JC, Hochstrasser DF, Williams KL, Gooley AA (1998) Extraction of membrane proteins by differential solubilization for separation using two-dimensional gel electrophoresis. Electrophoresis 19(5):837–844
50. Moratari SR, Saidelles APF, Barin JS, Flores EMM (2004) A simple procedure for decomposition of human hair using polypropylene vials to selenium determination by hydride generation atomic absorption spectrometry. Microchimica Acta 148:157–162
51. Moritz T, Johansson A (2007) Plant metabolomics. In: Griffiths W (ed) Metabolomics, metabonomics and metabolite profiling. RSC Publishing, Cambridge, pp 254–272
52. Mounicou S, Szpunar J, Lobinski R (2009) Metallomics: the concept and methodology. Chem Soc Rev 38:1119–1138
53. Oliver SG, Winson MK, Kell DB, Baganz F (1998) Systematic functional analysis of the yeast genome. Trends Biotechnol 16(9):373–378
54. Oniszczuka A, Olech M (2016) Optimization of ultrasound-assisted extraction and LC-ESI–MS/MS analysis of phenolic acids from *Brassica oleracea* L. var. *sabellica*. Ind Crop Prod 83:359–363
55. Pecsvaradi A, Nagy Z, Varga A, Vashegyi A, Labádi I, Galbács G, Zsoldos F (2009) Chloroplastic glutamine synthetase is activated by direct binding of aluminium. Physiol Plant 135:43–50
56. Peltier JB, Ytterberg AJ, Sun Q, Wijk KJ (2004) New functions of the thylakoid membrane proteome of *Arabidopsis thaliana* revealed by a simple, fast, and versatile fractionation strategy. J Biol Chem 279(47):49367–49383

57. Phizicky E, Bastiaens PIH, Zhu H, Snyder M, Fields S (2003) Protein analysis on a proteomic scale. Nature 422(6928):208–215
58. Raab A, Ploselli B, Munro C, Thomas-Oates J, Feldmann J (2009) Evaluation of gel electrophoresis conditions for the separation of metal-tagged proteins with subsequent laser ablation ICP-MS detection. Electrophoresis 30:303–314
59. Rakwal R, Komatsu S (2000) Role of jasmonate in the rice (Oryza sativa L.) self-defense mechanism using proteome analysis. Electrophoresis 21(12):2492–2500
60. Rep M, Dekker HL, Vossen JH, Boer AD, Houterman PM, Speijer D, Back JW, Koster CG, Cornelissen BJC (2002) Mass spectrometric identification of isoforms of PR proteins in xylem sap of fungus-infected tomato. Plant Physiol 130(2):904–917
61. Rojano-Delgado AM, Priego-Capote F, De Prado R, Castro MDL (2014) Qualitative/ quantitative strategy for the determination of glufosinate and metabolites in plants. Anal Bioanal Chem 406:611–620
62. Rostagno MA, Palma M, Barroso CG (2003) Ultrasound-assisted extraction of soy isoflavones. J Chromatogr A 1012:119–128
63. Salem MA, Jüppner J, Bajdzienko K, Giavalisco P (2016) Protocol: a fast, comprehensive and reproducible one-step extraction method for the rapid preparation of polar and semi-polar metabolites, lipids, proteins, starch and cell wall polymers from a single sample. Plant Methods 12(45):1–15
64. Santoni V, Kieffer S, Desclaux D, Masson F, Rabilloud T (2000) Membrane proteomics: use of additive main effects with multiplicative interaction model to classify plasma membrane proteins according to their solubility and electrophoretic properties. Electrophoresis 21(16):3329–3344
65. Sevcenco AM, Pinkse MWH, Bol E, Krijger GC, Wolterbeek HT, Verhaert PDE, Hagedoorn PL, Hagen WR (2009) The tungsten metallome of Pyrococcus furiosus. Metallomics 1:395–402
66. Sheffield J, Taylor N, Fauquet C, Chen S (2006) The cassava (Manihot esculenta Crantz) root proteome: protein identification and differential expression. Proteomics 6(5):1588–1598
67. Silva MO, Sussulini A, Arruda MAZ (2010) Metalloproteomics as na interdisciplinary area involving proteins and metals. Expert Rev Proteomics 7(3):387–400
68. Sun Y, Liu D, Chen J, Ye X, Yu D (2011) Effects of different factors of ultrasound treatment on the extraction yield of the all-trans-β-carotene from citrus peels. Ultrason Sonochem 18(1):243–249
69. Sussulini A, Garcia JS, Mesko MF, Flores EMM, Arruda MAZ (2006) Evaluation of soybean seed protein extraction focusing on metalloprotein analysis. Microchim Acta 158:173–180
70. Sussulini A, Garcia JS, Arruda MAZ (2007a) Microwave-assisted decomposition of polyacrylamide gels containing metalloproteins using mini-vials: an auxiliary strategy for metallomics studies. Anal Biochem 361:146–148
71. Sussulini A, Souza GHMF, Eberlin MN, Arruda MAZ (2007b) Comparative metallomics for transgenic and nontrasngenic soybeans. J Anal At Spectrom 22:1501–1506
72. Teo CC, Chong WPK, Ho YS (2013) Development and application of microwave-assisted extraction technique in biological sample preparation for small molecule analysis. Metabolomics 9:1109–1128
73. Vieira V, Prieto MA, Barros L, Coutinho JAP, Ferreira O, Ferreira ICFR (2017) Optimization and comparison of maceration and microwave extraction systems for the production of phenolic compounds from Juglans regia L. for the valorization of walnut leaves. Ind Crop Prod 107:341–352
74. Villas-Boas SG, Mas S, Åkesson M, Smedsgaard J, Nielsen J (2005) Mass spectrometry in metabolome analysis. Mass Spectrom Rev 24(5):613–646
75. Wang LJ, Weller CL (2006) Recent advances in extraction of nutraceuticals from plants. Trends Food Sci Technol 17(6):300–312
76. Watson BS, Asirvatham VS, Wang L, Sumner LW (2003) Mapping the proteome of barrel medic (Medicago truncatula). Plant Physiol 131(3):1104–1123

77. Wei J, Chen J, Liang X, Guo XJ (2016) Microwave-assisted extraction in combination with HPLC-UV for quantitative analysis of six bioactive oxoisoaporphine alkaloids in *Menispermum dauricum* DC. Biomed Chromatogr 30:241–248
78. Wu L, Song Y, Hu M, Yu C, Zhang H, Yu A, Ma Q, Wang Z (2015) Ionic-liquid-impregnated resin for the microwave-assisted solid-liquid extraction of triazine herbicides in honey. J Sep Sci 38(17):2953–2959
79. Xi J, Wang X, Li S, Zhou X, Yue L, Fan J, Hao D (2006) Polyethylene glycol fractionation improved detection of low-abundant proteins by two-dimensional electrophoresis analysis of plant proteome. Phytochemistry 67(21):2341–2348
80. Yang C, Wang J, Li D (2013) Microextraction techniques for the determination of volatile and semivolatile organic compounds from plants: a review. Anal Chim Acta 799:8–22
81. Zhang Y, Gao P, Xing Z, Jin S, Chen Z, Liu L, Constantino N, Wang X, Shi W, Yuan JS, Dai SY (2013) Application of an improved proteomics method for abundant protein cleanup: molecular and genomic mechanisms study in plant defense. Mol Cell Proteomics 12(11):3431–3442
82. Zhang H-F, Yang X-H, Wang Y (2011) Microwave assisted extraction of secondary metabolites from plants: current status and future directions. Trends Food Sci Technol 22:672–688

Chapter 8
Metaproteomics: Sample Preparation and Methodological Considerations

Benoit J. Kunath, Giusi Minniti, Morten Skaugen, Live H. Hagen,
Gustav Vaaje-Kolstad, Vincent G. H. Eijsink, Phil B. Pope,
and Magnus Ø. Arntzen

8.1 Introduction

High-throughput meta-omics methods hold great potential for microbial ecology and allow for unprecedented insight into organismal and functional composition of natural consortia in situ [1]. Collectively, the four omics techniques, genomics (DNA), transcriptomics (mRNA), proteomics (proteins), and metabolomics (metabolites), reflect the biology paradigm: genes for genetic information storage, transcripts for expression, proteins for structural and metabolic/enzymatic activities, and metabolites for substrates/inhibitors/products of metabolism [2]. Recent advances in -omics technologies have further extended their application to metagenomics [3], metatranscriptomics [4], metaproteomics [5], and meta-metabolomics [6]. Combined, these advances have revolutionized microbial ecology by providing scientists methods for addressing the complexity of microbial communities on a scale not attainable before.

The term metaproteomics was first coined in 2004 by Wilmes and Bond and describes the "large-scale characterization of the entire protein complement of environmental microbiota at a point in time" [5]. Since then, there has been a tremendous development in mass spectrometer speed, sensitivity, and accuracy, as well as concurrent development in software and wet-lab techniques. Metaproteomics today can identify thousands of proteins from an environmental sample and provide information on the actual proteins expressed by a microbial community, as well as

^{S}Authors "Benoit J. Kunath and Giusi Minniti" have contributed equally to this work.

B. J. Kunath · G. Minniti · M. Skaugen · L. H. Hagen · G. Vaaje-Kolstad · V. G. H. Eijsink
P. B. Pope · M. Ø. Arntzen (✉)
Faculty of Chemistry, Biotechnology and Food Science, Norwegian University of Life Sciences, Ås, Norway
e-mail: magnus.arntzen@nmbu.no

© Springer Nature Switzerland AG 2019
J.-L. Capelo-Martínez (ed.), *Emerging Sample Treatments in Proteomics*, Advances in Experimental Medicine and Biology 1073, https://doi.org/10.1007/978-3-030-12298-0_8

the abundances of these proteins and from which organisms they originated [7]. However, compared to the number of expected proteins present in such samples, and to identification rates observed in single-organism proteomics, metaproteomics is still facing significant obstacles in protein identification. Key challenges include (and are not limited to) protein extraction from difficult matrices, large sequence databases, sequence unavailability, and search engine sensitivity. This is evident from the large number of unknowns observed in metaproteomics data, i.e., many mass spectrometry MS/MS spectra that are not matched to a peptide sequence.

Here, we outline key methodological stages (Fig. 8.1) that are required for successful metaproteomic projects and highlight important on-going challenges. Further, we present two cases in which metagenomics and metaproteomics have

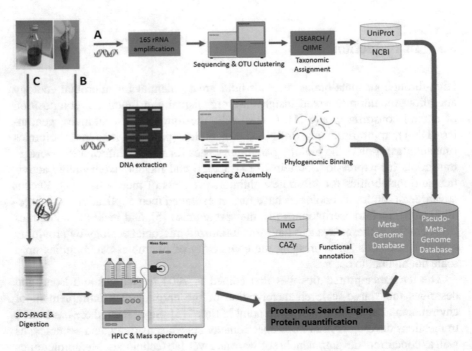

Fig. 8.1 Typical meta-omics workflows. (**A**) 16S rRNA analysis describes the structure of the microbial community and answers the question of "who is there." This gives the relative abundance of the different operational taxonomical units (OTUs) and can be used further to generate a sequence database for metaproteomics by querying public repositories for protein entries for relevant taxa. (**B**) Shotgun metagenomics generates sequence reads, which are subsequently assembled into contigs and phylogenomically binned into population-level genomes that represent a sample-specific database. Functional annotation from external databases such as the Integrated Microbial Genomes (IMG) or carbohydrate-active enzymes database (CAZy) can be queried to annotate the database. This describes the community potential and answers the question of "what is the community capable of doing." Furthermore, this approach provides a sample-specific sequence database for interpretation of metaproteomics data. (**C**) Mass spectrometry-based metaproteomics allows for detection of the actually expressed proteins and their relative quantification. This highlights the most abundant pathways in the microbial community and answers the questions "what is the community actually doing" and "who is doing what"

been combined toward a detailed understanding of the microbial community, its cooperative functions, and member contributions.

8.2 Sample Preparation and Protein Extraction

One of the most profound challenges in metaproteomics is rooted in the heterogeneity of habitats and matrices associated with the microbial community. In order to perform meaningful analyses, it is imperative that the proteins extracted from the environmental samples reflect the present microorganisms as correctly and impartial as possible. Hence, over the last couple of decades, a lot of work has been invested into developing robust protocols yielding unbiased, representative protein extracts from samples originating from habitats as different as the human gut [8], deep-sea thermal vents [9], activated sludge [10], or permafrost soil [11].

The first step in any environmental proteomics analysis is sample collection. In some cases, where the sample matrices are relatively homogenous and the abundance of microorganisms is high (e.g., activated sludge and animal gut/fecal samples), handling may be relatively straightforward. This is not the case for environments like oligotrophic seawater, where the original sample volume has to be reduced (by filtration and/or centrifugation) by orders of magnitude in order to yield a suitable concentration of microbes and hence proteins, to proceed with the analysis [12]. For most soil samples, consideration regarding the different soil strata needs to be accounted for, and sample homogenization (e.g., sieving to separate soil from leaf litter) is required before proceeding with cell disruption/lysis and protein extraction [13].

Many environmental samples may contain archaea, Gram-negative and Gram-positive bacteria, and fungi. In order to ensure proper disruption/lysis of all present cell types, as well as to detach cells from the surrounding matrix, a number of methods combining chemical and physical/mechanical treatment (reviewed in [13]) have been employed over the last few years. While differences dictated by the origin of the sample still exist, most current protocols share the use of lysis buffers containing moderate concentrations (0.1–5%) of detergent (usually SDS), in combination with heating and sonication and/or mechanical disruption (usually bead-beating). Variants of this treatment have been applied successfully to samples originating from gut [14], soil [11], acid mine drainage biofilms [15], and biogas reactors [7], and this treatment may therefore be considered a safe first approach to metaproteomics analysis of novel environmental samples.

In many cases, the proteins recovered by the above extraction procedure may be processed directly, following protocols like FASP [16] or STrap [17], where the proteins are digested on a solid support. Subsequently, the peptides are cleaned up by reversed-phase solid phase extraction (SPE), after which the peptide mixtures may be fractionated by high-performance liquid chromatography (HPLC) [17–19] prior to mass spectrometry (MS). In other cases, where the proteins have to be concentrated and/or separated from interfering substances from the sample matrix, a precipitation step may be required. Currently, the most common precipitation agent is trichloroacetic acid (TCA)/acetone, but precipitation using acetone, methanol/ammonium acetate, methanol/chloroform may also be used [20–22].

When working with samples containing significant amounts of humic substances (HS), sample preparation protocols often include a phenol extraction prior to protein precipitation [21, 23]. This procedure removes the majority of HS, which are known to interfere with downstream steps, including HPLC and MS [24–26]. While the phenol extraction step effectively removes the majority of the contaminating HS, it has been reported to result in poor protein recovery [20, 27, 28]. Recently, Qian and Hettich published an alternative and simple protocol for the removal of HS at the peptide level, by exploiting the differential solubility of peptides and humic acids at low pH [29].

For almost two decades, sample fractionation by one-dimensional (1D) gel electrophoresis followed by in-gel digestion and extraction of peptides has been a popular method to increase depth of analysis in proteomics experiments [30–32]. This method is simple and cheap, and has been shown to perform equally well or better when compared to other popular fractionation methods at the protein level [30, 32]. In a metaproteomics context, the added advantage of separating the proteins/peptides from interfering contaminants (which remain in the gel after processing) has been observed [33], and fractionation by 1D gel electrophoresis (isoelectric focusing or SDS-PAGE) is therefore recommended as a final step before further analysis of metaproteomics samples.

8.3 Sample Fractionation and HPLC-MS Analysis

Peptides generated by protein digestion are regularly further separated by HPLC prior to MS analysis. The most common separation technique is reversed- phase where peptides bind to a column containing a hydrophobic stationary phase (typically C18) and are eluted using a gradient of increasing amounts of an organic solvent such as acetonitrile. Separation can also be achieved using cation or anion exchange columns (SCX/WCX, SAX/WAX) or by reversed-phase C18 at low or high pH, to achieve deeper and multidimensional separation [34, 35]. Due to the complexity of microbial communities, large numbers of proteins are usually present in the samples and efficient separation of proteins and peptides is essential to achieve an adequate analysis depth. Kohrs et al. demonstrated this by comparing protein identification yields after three sample prefractionation steps and found that more prefractionation led to more proteins being identified, although at increased analysis time and cost [36]. The use of ultra-high performance liquid chromatography (UHPLC) and long columns (50 cm) allows excellent separation of many peptides present in the same sample, and gradients of up to 4 hours have been described [37]. Recently, even longer (75 cm) columns have become commercially available and are expected to improve peptide separation even further, potentially leading to more proteins being identified.

The sharp and narrow peaks delivered by a UHPLC system typically have an elution time between 4 and 20 seconds and thus require a mass spectrometer fast enough to be able to select and fragment every peptide precursor without too many peptides lost. The process of selecting a peptide and sequentially fragment it is referred to as data-dependent acquisition (DDA) and has been the most common way to operate a mass spectrometer in proteomics for years. DDA typically includes a strategy called TopN, where the N most intense peptides at any given time are selected for fragmentation and then the selected mass is excluded for 30–60 seconds to avoid duplicate sequencing. In very complex samples such as for metaproteomics, the DDA approach may not be optimal, as many peptides will not be among the TopN most intense peptides and thus not be selected for fragmentation. A promising new technique, called data-independent acquisition (DIA), does not select specific peptides but rather alternates between survey and fragmentation mode using a selected mass range of interest. This obviously breaks the association between each single peptide precursor and its fragment ions, and thus creates multiplexed MS/MS spectra, but also holds potential to overcome several of the DDA shortcomings [38]. It is expected that this technique finds wide applications in metaproteomics [1].

8.4 Sequence Databases

To allow the highest possible number of correctly identified proteins in a highly complex and heterogeneous microbial community, a comprehensive, thorough and high quality sequence database is required. Ideally, the database should be as dedicated as possible, i.e., containing the proteins potentially expressed by all the species present in the community, but nothing more. Using unnecessarily large databases, such as the complete bacterial section of UniProt (currently 58 M protein entries, August 2017), inevitably leads to a high fraction of false identifications (explained in Sect. 8.5.1). It is important to remember that, normally, only a minor fraction of the species present in the environmental sample has been sequenced and that only these (few) species are present in the public databases. Thus, public sequence databases are in most cases not comprehensive enough, which in turn lead to low identification rates, i.e., just a fraction of the MS/MS spectra is assigned a peptide sequence. The databases therefore have a pivotal role in the success of any metaproteomics experiment [39] and sample-specific databases, also called *matched databases*, generated using shotgun metagenomics data from the same sample, are essential.

Another middle-way option is using the public databases with a sample-specific filtering. Using 16S rRNA analysis, one can obtain the community structure and then construct a pseudo-metagenomic database by retrieving protein entries from defined taxa in public repositories (Fig. 8.1A), a strategy applied in Case Study 2 as well as in other studies [39, 40].

8.4.1 Database Generation from Metagenomics

A typical metagenomics analysis consists of five different steps: sequencing, assembly, binning, gene calling, and annotation. During these steps, a few requirements need to be met: First, the sequencing should attempt to cover the entire sample's complexity and diversity and avoid omitting rare populations present in the sample. The needed sequencing effort can be estimated using the community structure obtained by 16S rRNA analysis. Second, the assembly should generate long and accurate contigs since misassemblies may percolate into subsequent steps and likely result in errors during functional and/or taxonomic assignment. The longer and the more accurate the contigs are, the easier it is for binning software to assign them to a specific taxonomic bin. As populations of microorganisms from same species can differ from each other (strain-level variations), small sequence differences need to be captured in order to discriminate closely related individuals.

8.4.1.1 Sequencing

Numerous second-generation sequencing technologies now provide cheap, fast, and high-throughput sequencing [41]. Methods such as Illumina that produce short reads (between 150 and 300 bp) can generate a great amount of data resulting in a high sequencing depth and low error rates (<1%). However, assembly continuity can be relatively poor due to the limited ability of short reads to resolve long and repetitive sequences. The quality of the assembly can thus greatly be improved using longer reads (between 20 and 300 kb) such as provided by third-generation sequencing technologies like Oxford Nanopore's MinION and single-molecule real-time sequencing developed by Pacific Biosciences (PacBio). Although these techniques offer lower sequencing depth and higher error rates at the base level, the gain of long reads and thus better assembly continuity is favorable. Another interesting option is to include sequencing of mRNA; this will cover the expressed genes and therefore will allow to focus only on the active part of the community. mRNA sequencing may provide sequences lacking from the metagenomics, and its coassembly with DNA could potentially lead to improvement of the assembly continuity [42].

8.4.1.2 Assembly

Assembly is the key step to generate the large contigs required to maximize the completeness of predicted open reading frames (ORFs). A plethora of assemblers are currently available, such as IDBA_UD [43], MEGAHIT [44], and SPAdes [45]; for a recent review see [46]. These are designed to handle large datasets and highly uneven sequencing depth, which are typically characteristic of many metagenomics studies. The SPAdes assembly toolkit includes several algorithms, such as metaSPAdes and HybridSPAdes, which can accurately perform assembly by combining multiple libraries sequenced with various technologies (e.g., paired-reads and long error-prone reads). Even though hybrid assemblies (combining various sequencing technologies) are still infrequent, it has been shown that such an approach can improve the assembly continuity, the per-base accuracy and the assembly of universal marker genes, which improves the subsequent binning of metagenomics fragments and reconstruction of population-level genomes [47].

8.4.1.3 Binning

Binning is the postassembly assignment of contigs into genome bins and enables the study of individual microbial populations directly from metagenomes. The task of a binning tool is to assign an identifier to every assembled contig, with each identifier ideally corresponding to a single population-level genome. A multitude of software packages is currently available for this task and these are based on different approaches, the most important methodological difference being the supervised versus the unsupervised methods. Unsupervised software such as Metabat [48] and MaxBin [49] has the advantage that no training data is required to perform the classification. These methods capitalize on information from oligonucleotide composition, paired-end read linkage, mean contig coverage, per-sample coverage (when multiple metagenomes of the same community are generated), or combinations of these [50]. On the other hand, supervised software such as PhyloPythiaS+ [51], MEGAN [52], and taxator-tk [53] require a training phase and label contigs with taxa from an existing taxonomy database. Supervised classification usually allows more accurate discrimination than unsupervised procedures [54]. However, no single method will work for all samples, and as the field is developing, new software needs to be compared and evaluated. The "Critical Assessment of Metagenomic Information" (CAMI) [55] is such an initiative and aims to benchmark tools for binning, assembly and profiling on various datasets.

As means of validation, results for both assembly and binning should always be inspected to avoid misassemblies and incorrect assignments. For example, an abrupt change in GC% content and read coverage within the same contig can indicate a chimeric assembly. Changes in the contigs' characteristics can be visualized using different tools such as Anvi'o [56] or MGAviewer [57]. The quality of a bin can be defined by its level of completeness and contamination. These features can be estimated by the presence-absence of the universal genes (a set of genes present in every genome in only one copy) and the presence of multiple copies, respectively. The estimation can be performed either manually or by using automated methods, such as CheckM [58] or the HMM.essential.rb script of the enveomics collection [59]. A high-quality genome bin may be defined as being, at minimum, ≥90% complete and containing <5% of contamination. Furthermore, the genome bin should contain a 16S rRNA gene [60].

8.4.1.4 Gene Calling, Annotation, and Database Completion

Once a metagenomic dataset has been adequately assembled and taxonomically assigned, gene calling, also called ORF prediction, is required to identify protein-coding regions within the different bins. There are several methods for ORF prediction, including similarity-based and composition-based (also called ab initio) methods; for an overview, see [61]. Since metagenomic data from diverse communities tend to have low similarity with sequences from public databases, ab initio tools such as MetaGeneMark [62] have been preferred, and successfully used, in several studies [7, 63]. A new tool was recently developed using a graph-centric approach where the assembly graph produced by the assembler is used directly for predicting the ORFs [64]. This method, unlike previous contig-based ORF prediction approaches, captures the sequence variation present in the population and allows the prediction of more and longer ORFs. Once ORFs have been predicted, functional annotation of the genes is performed; this is elaborately explained in Sect. 8.7. Genome characterization can further be performed using Traitar [65] to accurately predict 67 diverse microbial phenotypes and thus aid in the understanding of what each organism is capable of doing. Finally, in the event that 16S rRNA analysis revealed community members that metagenomics were unable to detect, reference strains from public sequence repositories can be added to the metagenomics sequence database for completion [7, 63].

8.5 Protein Identification and Quantification in Metaproteomics

Compared to single-organism proteomics, metaproteomic researches are faced with a set of unique challenges regarding protein identification [66]. Microbial communities are highly complex and heterogeneous and may contain hundreds of different species, leading to very large and complex metagenomes typically encoding several hundred-thousand to a few million proteins. Further, microbial samples may contain many closely related proteins due to minor strain variations, horizontal gene transfer, or bacteriophage infections, as well as recurring functional domains, making peptide-to-protein-to-strain inference difficult [66, 67].

8.5.1 Protein Identification via Database Searching

Protein identification in proteomics can be achieved in (at least) three different ways where the most common approach is using a database and matches the experimental MS/MS spectra to theoretical fragmentation patterns of peptides found in an in silico digest of the database. To accomplish this task, a search engine is required, and there are a myriad to choose between: Mascot [68], Andromeda [69], X!Tandem [70], and many more; for a review, see [71]. The search engine outputs a list of identified peptide-to-spectrum matches (PSMs), a fraction of which will be false positives since the engine tries to match also "unmatchable" spectra to the database. The subsequent dilemma is then to distinguish between correct and incorrect peptide identifications and, therefore, strict control of the false discovery rate (FDR) is needed [72]. A common way to control the FDR is the target-decoy approach, where decoy sequences (reversed or scrambled protein sequences) are added to the database and matches to these decoys are regarded as known false positives [73]. The FDR is then estimated as the ratio of the number of decoy matches above a given threshold to the number of target database hits above the same threshold. This approach can then be used to limit the reported results to a given false-positive fraction, e.g., 1% FDR. For metaproteomics, which suffers from vastly increased search space compared to single-organism proteomics and a high similarity between sequences, it becomes more challenging to differentiate between a correct and an incorrect match, the FDR cannot be estimated reliably, and the target-decoy approach thus becomes less sensitive [2, 66]. In Case Study 1, our sample-specific sequence database had 607,516 protein entries, and led to an overall identification rate of 21% (the fraction of MS/MS spectra matched to a peptide sequence). This indicated that the target-decoy approach still worked with this database size, but likely sub-optimally. To increase the identification rate and regain some sensitivity of the analysis, authors have suggested to use a combination of more than one search engine [66] or to use a two-step database search, i.e., searching first a large database without decoys and from this then generate a refined smaller database for a second target-decoy search [74]. Efforts have been made to develop new search engines able to handle very large database sizes, as implemented in ComPIL [75] and Sipros [76], and newer alternatives include the use of machine learning decoy-free approaches, such as Nokoi [77], to model incorrect PSMs.

8.5.2 *Protein Identification via De Novo Sequencing*

Another promising identification approach, referred to as de novo sequencing, entails reading the amino acid sequence directly from the MS/MS spectra. The sequence of the peptide (or at least a short sequence tag) can then in turn be used in a sequence-similarity search, e.g., MS-BLAST (Basic Local Alignment Search Tool), to identify candidate proteins it originated from [78]. The benefit of this method is that it allows for protein identification via error-tolerant similarity searches rather than identity searches, as is the case for the database search approach described above, and therefore does not require a sequenced genome. Notably, when using this method, one may also discover unexpected post-translational modifications evident from the MS/MS spectra. The requirements, on the other hand, are that the quality of MS/MS spectra is high [79], and manual interpretation is needed due to the error-prone nature of de novo sequencing [66]. Several de novo sequencing software are available, including PepNovo+ [80], PEAKS [81], and more; for a recent review, see [82]. Muth et al. supplemented their metaproteomic analysis with peptide identifications from de novo sequencing and found that only 25% of the identified peptide sequences overlapped with peptides identified via the database approach [83]. Consequently, de novo-based identifications could be used as a complementary approach to add peptide identifications that are missed by database search engines.

8.5.3 Protein Identification via Spectral Libraries

A third way of identifying a peptide sequence is using spectral libraries. In this approach, experimental MS/MS spectra are compared with previously acquired and identified high-quality MS/MS spectra. Established spectral library search engines include X!Hunter [84], SpectraST [85], and BiblioSpec [86]. This is arguably the most limited method of the three, as one relies on the fact that the peptide has been identified by MS before. Relevant spectral libraries have to be available and this is difficult to obtain for metaproteomics as most libraries are generated from single-organism experiments [66].

8.5.4 Protein Quantification and Software Pipelines

Regarding the analysis of protein expression levels, label-free quantification (LFQ) has become the *modus operandi* in metaproteomics today as the technique can be applied directly to the protein identification data [87] without tedious and error-prone upstream labeling of proteins or peptides. Many different approaches and algorithms exist, but in general, they can be divided into MS1 methods, i.e., methods that use the intensity of the unfragmented peptide precursor as a quantitative measure, or MS2 methods that use the count of identified MS/MS spectra for a specific protein as a quantitative measure. MS1 methods are generally regarded to yield more accurate and robust measurements of the protein abundance as MS2 methods depend on peptide fragmentation, and thus on the less reproducible TopN selection process during DDA. The LFQ MS2 techniques are also referred to as spectral counting, and have been further developed into the exponentially modified protein abundance index (emPAI) [88] and the normalized spectral abundance factor (NSAF) [89]. A more comprehensive review of LFQ techniques can be found in [90].

Unfortunately, very few proteomic software packages have been designed with the "meta"-aspect in mind, and a fit-for-purpose modular software pipeline (including PSM identification with FDR control handling large databases, MS1 intensity-based quantification, visualization, and data storage) is still lacking in the field of metaproteomics. The MetaProteomeAnalyzer (MPA) [91] is a dedicated software aimed to tackle some of these challenges including the use of multiple search engines, two-step searching, grouping of similar proteins into "metaproteins," and powerful visualization of the results. MaxQuant [92] is currently one of the most used quantification software in proteomics and capable of raw data processing, protein identification via the build-in search engine Andromeda [69], and robust MS1 intensity-based label-free quantification (among others) with sophisticated algorithms for normalization across multiple raw files. However, software that are hard-coded to perform both protein identification and protein quantification may in fact be disadvantageous for metaproteomics due to the aforementioned underperformance of identification algorithms. A modular software design would be more flexible and allow exchanging, e.g., the search engine to a version optimized for large databases. Such a pipeline could be designed within the Galaxy for Proteomics (Galaxy-P) framework [93], while other promising pieces of a fit-for-purpose software include MoFF [94], a fast and robust MS1 intensity-based quantitative software. While in the absence of such a modular and optimized pipeline, in both Case Study 1 and Case Study 2, we have used MaxQuant for identification and quantification, particularly due to the robust MS1 quantification method.

8.6 Taxonomic Analysis

Taxonomic analysis aims to assign taxa from an existing taxonomy, such as the National Center for Biotechnology Information (NCBI) Taxonomy database, to an identified entity. This entity could be a peptide from a metaproteomics experiment or an assembled contig from a metagenomics experiment. By doing this, individual organisms (and their interactions) can be studied directly within a community. It is common to perform postassembly taxonomic assignment during the binning process in the metagenomics workflow, and this topic is elaborately covered in Sect. 8.4.1.3. This current section is therefore dedicated to taxonomic analysis using solely metaproteomics data and highlights unique taxonomical challenges related to protein inference.

8.6.1 Taxonomic Analysis from Metaproteomics Data

Taxonomic analysis based on metaproteomic datasets is tightly linked to protein inference; that is the nontrivial process of assigning a peptide to a protein [67]. Peptide sequences can be unique to a single protein, but often, and especially with short peptides, the peptide sequence may match to several proteins with a similar sequence. A peptide's uniqueness is partly linked to the peptide length, as longer peptides are more likely to be unique while short peptides often match so many proteins that they are essentially useless. The mass spectrometer can therefore advantageously be tuned to disregard short peptides and gain more analysis time for longer peptides.

Proteomic search engines will not only produce a list of identified peptides, but also assemble the peptides into proteins, or rather protein groups, where several proteins are grouped together as the underlying peptides cannot distinguish the proteins adequately. In metaproteomics, it may also occur that the proteins making up one such protein group individually belong to different species; at this point, it becomes difficult to assign taxa to this protein group. Hence, there is a tight link between protein inference, protein grouping and taxonomic assignment. One strategy to tackle this challenge is to take advantage of a concept from graph theory called the lowest common ancestor (LCA), where the LCA refers to the most specific node in a tree that contains all descendants under question. Transferring this conception to taxonomy, one can say that the LCA of a protein group would be the most specific taxonomical rank that explains the origin of all the proteins in the group. The MetaProteomeAnalyzer generates protein groups called meta-proteins and they can be assigned to the LCA of the proteins making up the protein group. Others have suggested weighting LCA calculations based on peptide abundance [95], while another option is to limit the resolution of the analysis to, e.g., the genus level. Which strategy to follow is in the end a question of how deep one needs to go in the taxonomical analysis to answer the biological question at hand, and how deep one can go based on the experimental data. Population-level metaproteomics may be very relevant in some cases, e.g., when homologous strains have different functional traits, while in other scenarios, the genus level might be more appropriate as long as a function can be inferred. As described in Case Study 1, Hagen et al. [7] used functional annotations as guidelines in the protein inference and taxonomical assignment. In brief, if the proteins in a protein group could be assigned to multiple species, the pathway context was used for guidance, i.e., the protein group was assigned to the species that had the highest number of other protein identifications supporting the same pathway. Other authors have suggested a workflow using peptide sequences for BLAST analyses, followed by taxonomic analysis of the results using the homology-matching software MEGAN [52, 66]. Another tool for taxonomic analysis of metaproteomic data is UniPept [96]. This tool enables retrieval of taxa and LCA for a set of peptides and provides powerful, interactive visualization of the biodiversity. In this approach, the UniProt knowledgebase [97] is used as data source, but the web server also allows for uploading custom genomes to enhance the analysis. Other tools available for taxonomic analysis of metaproteomics data include, among others, Prophane [98] and Pipasic [99]; a recent review of these and additional data analysis tools in metaproteomics can be found in [100].

8.7 Functional Annotation and Analysis

In order to generate insight into ongoing biological processes and to know "who is doing what" in the microbial community, functional annotation of the proteins is necessary. In brief, this is accomplished by querying (several) public databases to retrieve functional tags for each of the proteins and then functionally group these proteins into, e.g., metabolic pathways. Quantitative expression patterns can then be correlated with this functional information with tools like Perseus [101] and Pathview [102] to generate functionally decorated heat maps or pathways with over-laying expression data. The annotation process can be performed during database generation, typically using annotation pipelines from the metagenomics field, or it can be performed at the protein level, before or after, metaproteomics analysis.

8.7.1 Annotation Databases

There are multiple approaches to functionally annotate proteins, including homology-based methods such as BLAST [103] and the use of hidden Markov models in HMMER [104]; motif- and pattern-based methods, such as sequence signature searches; and context-based methods such as genomic neighborhood analysis [105, 106]. Numerous tools and databases are publicly available, including, among others, the UniProt knowledgebase [97], COG (Clusters of Orthologous Groups) for functional grouping [107], Pfam (the protein families database) for identification of protein families and domains [108], TIGRfam for full-length protein families [109], and Enzyme Commission (EC) numbers for numerical classification of enzymes based on the chemical reactions they catalyze [110]. In addition to these, specific databases enable reconstruction of pathway maps for cellular and organismal functions. Key examples are the Kyoto Encyclopedia for Genes and Genomes (KEGG) [111] and the MetaCyc (metabolic Pathway Database) [112]. In Case Study 1, we used pathways retrieved from KEGG and reconstructed genomes, and decorated these with protein abundance values from metaproteomics to highlight abundant, and likely most active, pathways in the microbial community.

The Gene Ontology (GO) is a structured and precisely defined, controlled vocabulary for describing the roles of genes and gene products and provides three so-called ontologies that describe (1) molecular functions, (2) biological processes, and (3) cellular components independent of species [113]. GO is commonly used in single-organism proteomics for enrichment analysis, where a list of identified proteins (e.g., upregulated proteins) is compared versus, e.g., the whole predicted proteome of the organism, and statistically overrepresented GO terms in the list are highlighted, analogous to gene set enrichment analysis in functional genomics [114]. Several tools are available for such analysis, including DAVID [115] and g:Profiler [116], where the latter also supports sorted lists, e.g., lists sorted by protein abundance. However, these tools require a single organism as background and currently do not support analysis of metaproteomic datasets. However, the GO terms can still be used for functional classification. In Case Study 2, we utilized GO terms to functionally annotate and classify proteins produced by the bacterial community residing in salmon skin mucus.

InterPro is a tool for predicting domains and important sites in proteins and capitalizes on integrating information from multiple databases with individual strengths, into a single searchable resource, reducing redundancy and making the interpretation easier [117]. InterPro can also predict GO annotations for proteins based on functional domains detected, and thus constitutes a link between sequence signatures and the GO hierarchy. While the web service is only able to predict domains on single protein entries, the InterProScan [118] is a stand-alone resource that can be used for analyzing multi-FASTA files against InterPro signatures. Similarly, a number of software pipelines from the metagenomics field are available. The Integrated Microbial Genomes with Microbiome Samples-Expert Review (IMG/MER) system provides comprehensive functional annotation, functional profiling, and curation of large metagenomics datasets [119]. The newest developments from KEGG to this end include BlastKOALA and GhostKOALA, tools which are able to map multi-FASTA files to KEGG's functional ontology and to pathway- and module maps [120].

8.7.2   Specialized Databases for Annotation

In addition to the abovementioned general functional prediction tools, databases specialized in certain fields can also be queried. The carbohydrate-active enzymes database (CAZy) is one such database and specialized in display and analysis of genomic, structural, and biochemical information on carbohydrate-active enzymes [121]. CAZy contains more than 300 families of catalytic and ancillary modules, classified as glycoside hydrolases (GHs), glycosyltranferases (GTs), polysaccharide lyases (PLs), carbohydrate esterases (CEs), auxiliary activities (AAs), and carbohydrate binding modules (CBMs). The CAZy group actively develops tools for unambiguous high-throughput modular and functional annotation of CAZymes in sequences issued from genomic and metagenomic efforts. In addition to CAZy, alternative efforts exist for automated CAZyme annotation, including dbCAN [122] and CAT [123]. MEROPS is another specialized database focusing on proteolytic enzymes, their substrates, and their inhibitors [124] and hierarchically group enzymes into species, families, and clans.

8.7.3   Limitations of In Silico Functional Annotations

It is important to realize that functional predictions and domain identifications do not necessarily display the complete picture. Predictions might be incomplete (e.g., unrecognized sequence stretches or domains of unknown functions) or even erroneous. Furthermore, many proteins are built up in a multimodular fashion, and some may exhibit more than one function, as in the so-called moonlighting proteins [125]. During the annotation process, these proteins will attract several, and sometimes even conflicting, functional tags, which may preclude correct interpretation. In silico functional predictions should ideally be validated by functional studies such as protein purification and thorough characterization, but this represents a large undertaking and is often outside the scope of metaproteomic experiments. However, functional metaproteomic experiments can be performed. For example, by using ^{15}N-based stable isotope labeling in mammals (SILAM) and application of activity-based probes for protein enrichment, Mayers et al. were able to quantify mouse and microbial proteins with specific functionalities using metaproteomics [126]. A related technique for looking at metabolic function is protein-based stable isotope probing (protein-SIP) where ^{13}C- or ^{15}N-labeled substrates are used to retrieve information on which members of the bacterial community are responsible for the metabolism of a particular substrate [127].

8.8 Selected Examples

8.8.1 Case Study 1: Biogas Reactors

As a response to environmental challenges, the world is facing a transition toward more efficient and clever utilization of renewable biological resources. Organic waste from industry, agriculture, aquaculture, and households is a major concern as the traditional waste treatments usually involve incineration or landfilling and emission of greenhouse gases. A viable alternative is the utilization of biological-mediated processes to convert bio-waste to methane or "biogas" as an approach for sustainable waste disposal while producing renewable energy. This process, hereby referred to as Anaerobic Digestion (AD), is carried out by a collection of microbes collaborating to degrade the organic compounds through tight and complex food webs. A comprehensive understanding of this microbial community is crucial for effective biogas production and most research initiatives addressing this matter have so far focused on the presence/absence of microbial populations and associated genes through targeted amplicon sequencing (16S rRNA gene sequencing) and shotgun metagenomics sequencing approaches [128]. Although providing valuable information of microbial communities in such ecosystems, the genetic content only reveals potential features, and does not necessarily correlate with the actual activity and interactions between community members. Combined meta-omics approaches have the potential to overcome these limitations and have been successful in revealing community function and metabolic activity in a number of ecological studies [11, 129].

In two recent studies reported by our group [7, 63], AD processes in industrial scale biogas reactors were investigated using metagenomics combined with metaproteomics, essentially as depicted in Fig. 8.1B, C. Both reactors are part of larger plants in Norway that treat waste from nearby industrial and municipal households, and the biogas, a mixture of primarily methane and CO_2, is further upgraded to biofuels (i.e., methane), to be used in vehicles and buses in the region. The overreaching objective in these two studies was to increase the understanding of the actual metabolism occurring within industrial scale reactors. Population genomes mined from metagenomics dataset were examined at the genetic level, and functional annotations of the genes from IMG provided insight into the metabolic potential. By using this genomic information as sample-specific databases for the metaproteomics datasets, expressed genes were mapped back to the microbial populations at species level. As for the metaproteome, a protein had to be present in both of a duplicated proteomics dataset to be considered valid, and a pathway-guided taxonomical assignment was applied. In brief, if the proteins in a protein group could be mapped to two different species and the protein belonged to a pathway, it was assigned to the species that had the highest number of other protein identifications supporting the same pathway. MaxQuant allowed for quantification of the proteins, which was used as an indicator of activity. Overall, this approach made it possible to draft a reconstruction of key metabolic pathways (Fig. 8.2) and link novel microbial populations to steps in the AD process (Fig. 8.3).

In the first report [63], the AD examined was operating under mesophilic conditions (37 °C), while the other case [7] concerned AD at higher temperature (60 °C). Common for both was that food waste accounted for a major portion of the feedstock. Although food waste holds a large biogas potential when used as substrate for AD, this feedstock is also associated with a higher risk of process failures due to accumulation of ammonia, a key intermediate metabolite from protein degradation [130]. When reaching toxic levels, free ammonia inhibits the acetoclastic methanogens responsible for the final step in the AD process, where acetate is converted to methane [131, 132]. Thus, AD of proteinaceous material often promotes a two-step mechanism as an alternative acetate conversion route, in which syntrophic acetate-oxidizing bacteria (SAOB) have a pivotal role by oxidizing acetate to CO_2 and H_2 [131], which are further converted to methane by a group of hydrogen consuming methanogens. This syntrophic association with a hydrogen scavenger is essential, as the oxidation of acetate is thermodynamically unfavorable ($\Delta G > 0$) under anaerobic conditions unless the hydrogen concentration is kept at a minimum [133]. Despite the importance of these cooperative interactions in biogas processes, only a few cultivable SAOBs have been recovered and described [134–136]. In our biogas reactor studies, we detected a novel *Firmicutes* species with the highest abundance of Wood-Ljungdahl associated enzymes in the mesophilic community, indicating the presence of highly active populations of novel SAOBs (Fig. 8.2). Moreover, a population consisting of at least two strains of putative novel SAOBs was reported in the thermophilic reactor [7] where the metaproteomics-driven metabolic reconstruction also suggested SAOBs with a broader metabolic diversity than previously detected. These two cases illustrate how a combination of two meta-omics strategies can provide additional depth when exploring the functionality of microbial communities.

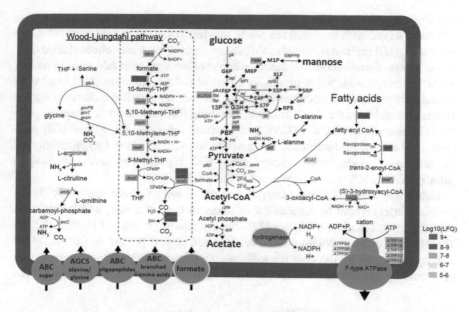

Fig. 8.2 Selected metabolic features of a novel syntrophic acetate oxidizing bacterium (SAOB) as inferred from the genome and proteome data. Metaproteomic analysis, using MaxQuant label-free quantification (LFQ) values (scaled as indicated at the bottom right), allowed us to annotate the pathways with expression values and showed all Wood-Ljungdahl pathway enzymes being highly abundant, as well as enzymes associated with fatty acid degradation. In particular, the high abundance of a unidirectional fructose 1,6-bisphosphate aldolase/phosphatase (ALDOA) supported the hypotheses that this bacterium performs gluconeogenesis (anabolic glucose metabolism). (The figure is reproduced from [63] with permission from American Society for Microbiology)

8.8.2   Case Study 2: Salmon Skin Microbiome

Norway is the world's largest producer and exporter of Atlantic salmon (*Salmo salar*). Farmed fish are often exposed to stress conditions (handling, netting, sorting, transport, high stocking density) and are thus more susceptible to diseases compared to wild fish [137], causing economic losses in the aquaculture sector. Skin disorders are one of the main problems associated with fish mortality in aquaculture, and it is estimated that 1.1–2.5% of farmed fish die due to skin ulceration [138]. The skin of the teleost is coated with mucus, which contains mucins (high O-glycosylated proteins) and several antimicrobial compounds, such as immunoglobulins, antimicrobial peptides, and various enzymes, to protect the skin from the external environment [139]. Despite the protective function of mucus, some microorganisms are able to colonize the mucosal surfaces and the carbon source provided by mucins may aid in the colonization process [140]. The complex interactions that likely exist between fish mucus and its bacterial community are not well understood.

As part of an effort to understand these interactions, we used a metaproteomic approach to investigate the proteins present in the mucus of farmed *Salmo salar*, focusing on both host and bacterial proteins. Mucus from three farmed salmon was collected, mixed with sterilized sea water, and incubated for 9 days at 10 °C using three biological replicates. Samples for proteomic analysis were collected after 0, 2, 5, and 9 days of incubation. The main challenge of working with salmon skin mucus is the viscosity [141], which precludes an initial 0.2 μm filtration steps needed to collect only the secreted fraction of proteins. Instead, we centrifuged at 5500 × *g* for 10 minutes and used the liquid fraction, assuming this fraction to be (largely) cell-free. Trichloroacetic acid (TCA) precipitation followed by SDS-PAGE was performed as sample cleanup and fractionation steps prior to in-gel trypsinization of the proteins. Peptides were desalted using C18 ZipTips and analyzed using a nanoHPLC-MS/MS system (explained in Sect. 8.3).

Using the whole UniProt bacterial section as search space (currently 58 M protein entries) would be impractical in terms of time and size, and it would be challenging to calculate a correct FDR for the results (explained in Sect. 8.5.1). Therefore, in order to obtain a more detailed and correct picture of the biological activity employed by the host microbiome, a pseudo-metagenomic database was generated by extracting and concatenating UniProt and/or NCBI protein entries

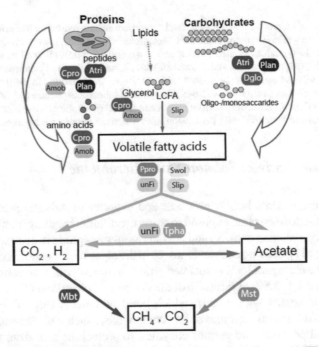

Fig. 8.3 Hypothetical model of "who does what," i.e., the carbon flux, in the FREVAR biogas reactor presented in [7]. The functional roles of dominant phylotypes were inferred based on the expressed proteins detected in the metaproteomics datasets. The colored arrows indicate key metabolic pathways: acetogenesis (pink), methanogenesis (blue), and syntrophic metabolic processes (green). The syntrophic acetate-oxidizing bacteria (SAOB) is abbreviated "unFi." See Table 2 in [7] for an explanation of organism abbreviations. (The figure is reproduced with permission from American Society for Microbiology)

from the genera observed by 16S rRNA gene sequencing analysis, a strategy also suggested by other authors [39, 40]. To generate the pseudo-metagenomic database, we followed the protocol described in Fig. 8.1a. In brief, DNA was extracted at day 9 and the genes coding for the bacterial 16S rRNA were amplified and sequenced by Miseq (Illumina). Usearch [142] within QIIME [143] was used to analyze the raw data output, and only the most abundant taxa (operational taxonomical units (OTUs) >0.2% of the total number of sequences; 18 bacterial genera) were used for the construction of the database. In addition, the UniProt entries for *Salmo salar* were added in order to detect host proteins. The final database used for proteomics had 2,224,787 protein entries in total. MaxQuant [92] was used to analyze the MS raw data and the LFQ values reported by MaxQuant were utilized to calculate summed protein abundances at genus level (Fig. 8.4). In addition, Gene Ontology (GO) was used for a functional annotation of the bacterial and salmon skin mucus proteins (explained in Sect. 8.7.1). The results of this study showed a high abundance of *S. salar* proteins at days 0 and 2, which decreased at later time points as bacterial proteins increased. GO analysis revealed a high expression of proteases and taken together with the observed decrease of *S. salar* proteins and increase of bacterial proteins, this could suggest that some of the bacteria may have the capacity to degrade the proteins residing in the salmon mucus. In overall, *Vibrio* was the most abundant genus observed.

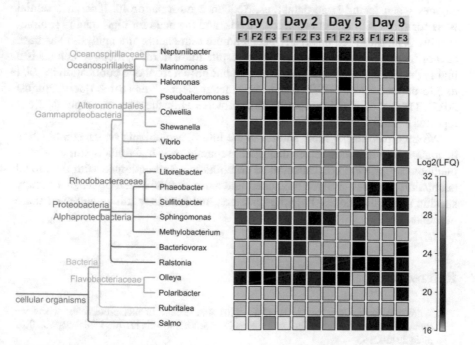

Fig. 8.4 Heat map indicating the summed protein abundances in the salmon mucus during 9 days of incubation. Each row represent bacterial genus, and *S. salar* is shown in bottom row (Unpublished data)

It should be noted that sample preparation is of particular importance when studying a fish skin proteome. One may dissolve mucus in chaotropic agents or detergents to solubilize insoluble mucus proteins (sometimes aided by sonication), or one may only analyze the soluble mucus fraction obtained by centrifugation (like in the example described above). Both sample preparation methods are useful, but the former may give a high background of cytoplasmic proteins since skin mucus also is known to contain immune cells, keratocytes, and shed epithelial cells that may lyse when exposed to chaotropic agents, detergents, and/or sonication. Indeed, existing proteomic studies of fish skin mucus report the finding of salmon proteins that typically locate to the cytoplasm of cells [144, 145]. Thus, for interpretation of mucus proteome data, both sample treatment and downstream analysis must be conducted carefully in order to prevent misinterpretations.

8.9   Summary and Outlook

Given the intricate complexity found in microbial communities and the large differences in protein abundances between genera, metaproteomic projects still face many challenges. The most prominent of these are perhaps unbiased protein extraction during sample preparation and the underperformance of protein search engines when facing large databases. Sequence information itself is an essential issue for successful metaproteomic analysis, and the need for high-quality sample-specific databases (i.e., upstream metagenomic sequencing and analysis) has been acknowledged by the community [39]. Despite these hurdles, metaproteomics is a highly promising, emerging technique that has grown from one publication in 2004 to 336 publications by 2016 (PubMed; "metaproteom*" as search query; August 2017). The first publication reported three proteins while the most recent publications reported several thousand proteins.

We anticipate seeing more studies in the future, capitalizing on the combination of meta-omics methods rather than their separate use. Individually, these -omics methods can provide great insight; in combination, and equipped with functional annotation, they can provide a detailed understanding of which organisms occupy specific metabolic niches, how they interact, and how they utilize and even share nutrients.

References

1. Wilmes P, Heintz-Buschart A, Bond PL (2015) A decade of metaproteomics: where we stand and what the future holds. Proteomics 15(20):3409–3417. https://doi.org/10.1002/pmic.201500183
2. Xiong W, Abraham PE, Li Z, Pan C, Hettich RL (2015) Microbial metaproteomics for characterizing the range of metabolic functions and activities of human gut microbiota. Proteomics 15(20):3424–3438. https://doi.org/10.1002/pmic.201400571

3. Tyson GW, Chapman J, Hugenholtz P, Allen EE, Ram RJ, Richardson PM, Solovyev VV, Rubin EM, Rokhsar DS, Banfield JF (2004) Community structure and metabolism through reconstruction of microbial genomes from the environment. Nature 428(6978):37–43. https://doi.org/10.1038/nature02340

4. Aliaga Goltsman DS, Comolli LR, Thomas BC, Banfield JF (2015) Community transcriptomics reveals unexpected high microbial diversity in acidophilic biofilm communities. ISME J 9(4):1014–1023. https://doi.org/10.1038/ismej.2014.200

5. Wilmes P, Bond PL (2004) The application of two-dimensional polyacrylamide gel electrophoresis and downstream analyses to a mixed community of prokaryotic microorganisms. Environ Microbiol 6(9):911–920. https://doi.org/10.1111/j.1462-2920.2004.00687.x

6. Dettmer K, Aronov PA, Hammock BD (2007) Mass spectrometry-based metabolomics. Mass Spectrom Rev 26(1):51–78. https://doi.org/10.1002/mas.20108

7. Hagen LH, Frank JA, Zamanzadeh M, Eijsink VG, Pope PB, Horn SJ, Arntzen MO (2017) Quantitative metaproteomics highlight the metabolic contributions of uncultured phylotypes in a thermophilic anaerobic digester. Appl Environ Microbiol 83(2). https://doi.org/10.1128/AEM.01955-16

8. Lee PY, Chin S-F, Neoh H-M, Jamal R (2017) Metaproteomic analysis of human gut microbiota: where are we heading? J Biomed Sci 24:36. https://doi.org/10.1186/s12929-017-0342-z

9. Urich T, Lanzén A, Stokke R, Pedersen RB, Bayer C, Thorseth IH, Schleper C, Steen IH, Øvreas L (2014) Microbial community structure and functioning in marine sediments associated with diffuse hydrothermal venting assessed by integrated meta-omics. Environ Microbiol 16:2699–2710. https://doi.org/10.1111/1462-2920.12283

10. Püttker S, Kohrs F, Benndorf D, Heyer R, Rapp E, Reichl U (2015) Metaproteomics of activated sludge from a wastewater treatment plant – a pilot study. Proteomics 15:3596–3601. https://doi.org/10.1002/pmic.201400559

11. Hultman J, Waldrop MP, Mackelprang R, David MM, McFarland J, Blazewicz SJ, Harden J, Turetsky MR, McGuire AD, Shah MB, VerBerkmoes NC, Lee LH, Mavrommatis K, Jansson JK (2015) Multi-omics of permafrost, active layer and thermokarst bog soil microbiomes. Nature 521:208–212. https://doi.org/10.1038/nature14238

12. Wang D-Z, Xie Z-X, Zhang S-F (2014) Marine metaproteomics: current status and future directions. J Proteomics 97:27–35. https://doi.org/10.1016/j.jprot.2013.08.024

13. Keiblinger KM, Fuchs S, Zechmeister-Boltenstern S, Riedel K (2016) Soil and leaf litter metaproteomics – a brief guideline from sampling to understanding. FEMS Microbiol Ecol 92. https://doi.org/10.1093/femsec/iw180

14. Zhang X, Li L, Mayne J, Ning Z, Stintzi A, Figeys D (2017) Assessing the impact of protein extraction methods for human gut metaproteomics. J Proteomics 180:120. https://doi.org/10.1016/j.jprot.2017.07.001

15. Méndez-García C, Peláez AI, Mesa V, Sánchez J, Golyshina OV, Ferrer M (2015) Microbial diversity and metabolic networks in acid mine drainage habitats. Front Microbiol 6:475. https://doi.org/10.3389/fmicb.2015.00475

16. Wiśniewski JR, Zougman A, Mann M (2009) Combination of FASP and StageTip-based fractionation allows in-depth analysis of the hippocampal membrane proteome. J Proteome Res 8:5674–5678. https://doi.org/10.1021/pr900748n

17. Zougman A, Selby PJ, Banks RE (2014) Suspension trapping (STrap) sample preparation method for bottom-up proteomics analysis. Proteomics 14:1006–1000. https://doi.org/10.1002/pmic.201300553

18. Hernandez-Valladares M, Aasebø E, Mjaavatten O, Vaudel M, Bruserud Ø, Berven F, Selheim F (2016) Reliable FASP-based procedures for optimal quantitative proteomic and phosphoproteomic analysis on samples from acute myeloid leukemia patients. Biol Proced Online 18:13. https://doi.org/10.1186/s12575-016-0043-0

19. Wisniewski JR, Zougman A, Mann M (2009) Combination of FASP and StageTip-based fractionation allows in-depth analysis of the hippocampal membrane proteome. J Proteome Res 8(12):5674–5678. https://doi.org/10.1021/pr900748n

20. Keiblinger KM, Wilhartitz IC, Schneider T, Roschitzki B, Schmid E, Eberl L, Riedel K, Zechmeister-Boltenstern S (2012) Soil metaproteomics – comparative evaluation of protein extraction protocols. Soil Biol Biochem 54:14–24. https://doi.org/10.1016/j.soilbio.2012.05.014

21. Speda J, Johansson MA, Carlsson U, Karlsson M (2017) Assessment of sample preparation methods for metaproteomics of extracellular proteins. Anal Biochem 516:23–36. https://doi.org/10.1016/j.ab.2016.10.008

22. Wessel D, Flugge UI (1984) A method for the quantitative recovery of protein in dilute solution in the presence of detergents and lipids. Anal Biochem 138(1):141–143

23. Wang W, Vignani R, Scali M, Cresti M (2006) A universal and rapid protocol for protein extraction from recalcitrant plant tissues for proteomic analysis. Electrophoresis 27:2782–2786. https://doi.org/10.1002/elps.200500722

24. Arenella M, D'Acqui LP, Pucci A, Giagnoni L, Nannipieri P, Renella G (2014) Contact with soil-borne humic substances interfere with the prion identification by mass spectrometry. Biol Fertil Soils 50:1009–1013. https://doi.org/10.1007/s00374-014-0922-y

25. Arenella M, Giagnoni L, Masciandaro G, Ceccanti B, Nannipieri P, Renella G (2014) Interactions between proteins and humic substances affect protein identification by mass spectrometry. Biol Fertil Soils 50:447–454. https://doi.org/10.1007/s00374-013-0860-0

26. Piccolo A, Spiteller M (2003) Electrospray ionization mass spectrometry of terrestrial humic substances and their size fractions. Anal Bioanal Chem 377:1047–1059. https://doi.org/10.1007/s00216-003-2186-5

27. Benndorf D, Balcke GU, Harms H, von Bergen M (2007) Functional metaproteome analysis of protein extracts from contaminated soil and groundwater. ISME J 1:224–234. https://doi.org/10.1038/ismej.2007.39

28. Giagnoni L, Magherini F, Landi L, Taghavi S, Modesti A, Bini L, Nannipieri P, Van der lelie D, Renella G (2011) Extraction of microbial proteome from soil: potential and limitations assessed through a model study. Eur J Soil Sci 62:74–81. https://doi.org/10.1111/j.1365-2389.2010.01322.x

29. Qian C, Hettich RL (2017) Optimized extraction method to remove humic acid interferences from soil samples prior to microbial proteome measurements. J Proteome Res 16:2537–2546. https://doi.org/10.1021/acs.jproteome.7b00103

30. Dowell JA, Frost DC, Zhang J, Li L (2008) Comparison of two-dimensional fractionation techniques for shotgun proteomics. Anal Chem 80:6715–6723. https://doi.org/10.1021/ac8007994

31. Lee C-L, Hsiao H-H, Lin C-W, Wu S-P, Huang S-Y, Wu C-Y, Wang AH-J, Khoo K-H (2003) Strategic shotgun proteomics approach for efficient construction of an expression map of targeted protein families in hepatoma cell lines. Proteomics 3:2472–2486. https://doi.org/10.1002/pmic.200300586

32. Weston LA, Bauer KM, Hummon AB (2013) Comparison of bottom-up proteomic approaches for LC-MS analysis of complex proteomes. Anal Methods 5:4615. https://doi.org/10.1039/C3AY40853A

33. Heyer R, Kohrs F, Reichl U, Benndorf D (2015) Metaproteomics of complex microbial communities in biogas plants. J Microbial Biotechnol 8:749–763. https://doi.org/10.1111/1751-7915.12276

34. Yang F, Shen Y, Camp DG 2nd, Smith RD (2012) High-pH reversed-phase chromatography with fraction concatenation for 2D proteomic analysis. Expert Rev Proteomics 9(2):129–134. https://doi.org/10.1586/epr.12.15

35. Zhang X, Fang A, Riley CP, Wang M, Regnier FE, Buck C (2010) Multi-dimensional liquid chromatography in proteomics – a review. Anal Chim Acta 664(2):101–113. https://doi.org/10.1016/j.aca.2010.02.001

36. Kohrs F, Heyer R, Magnussen A, Benndorf D, Muth T, Behne A, Rapp E, Kausmann R, Heiermann M, Klocke M, Reichl U (2014) Sample prefractionation with liquid isoelectric focusing enables in depth microbial metaproteome analysis of mesophilic and thermophilic biogas plants. Anaerobe 29:59–67. https://doi.org/10.1016/j.anaerobe.2013.11.009

37. Pirmoradian M, Budamgunta H, Chingin K, Zhang B, Astorga-Wells J, Zubarev RA (2013) Rapid and deep human proteome analysis by single-dimension shotgun proteomics. Mol Cell Proteomics 12(11):3330–3338. https://doi.org/10.1074/mcp.O113.028787
38. Bilbao A, Varesio E, Luban J, Strambio-De-Castillia C, Hopfgartner G, Muller M, Lisacek F (2015) Processing strategies and software solutions for data-independent acquisition in mass spectrometry. Proteomics 15(5–6):964–980. https://doi.org/10.1002/pmic.201400323
39. Tanca A, Palomba A, Fraumene C, Pagnozzi D, Manghina V, Deligios M, Muth T, Rapp E, Martens L, Addis MF (2016) The impact of sequence database choice on metaproteomic results in gut microbiota studies. Microbiome 4(1):51
40. Muth T, Renard BY, Martens L (2016) Metaproteomic data analysis at a glance: advances in computational microbial community proteomics. Expert Rev Proteomics 13(8):757–769. https://doi.org/10.1080/14789450.2016.1209418
41. Bragg L, Tyson GW (2014) Metagenomics using next-generation sequencing. In: Paulsen IT, Holmes AJ (eds) Environmental microbiology: methods and protocols. Humana Press, Totowa, pp 183–201. https://doi.org/10.1007/978-1-62703-712-9_15
42. Heintz-Buschart A, May P, Laczny CC, Lebrun LA, Bellora C, Krishna A, Wampach L, Schneider JG, Hogan A, de Beaufort C, Wilmes P (2016) Integrated multi-omics of the human gut microbiome in a case study of familial type 1 diabetes. Nat Microbiol 2:16180. https://doi.org/10.1038/nmicrobiol.2016.180
43. Peng Y, Leung HC, Yiu SM, Chin FY (2012) IDBA-UD: a de novo assembler for single-cell and metagenomic sequencing data with highly uneven depth. Bioinformatics 28(11):1420–1428. https://doi.org/10.1093/bioinformatics/bts174
44. Li D, Liu C-M, Luo R, Sadakane K, Lam T-W (2015) MEGAHIT: an ultra-fast single-node solution for large and complex metagenomics assembly via succinct de Bruijn graph. Bioinformatics 31(10):1674–1676. https://doi.org/10.1093/bioinformatics/btv033
45. Bankevich A, Nurk S, Antipov D, Gurevich AA, Dvorkin M, Kulikov AS, Lesin VM, Nikolenko SI, Pham S, Prjibelski AD, Pyshkin AV, Sirotkin AV, Vyahhi N, Tesler G, Alekseyev MA, Pevzner PA (2012) SPAdes: a new genome assembly algorithm and its applications to single-cell sequencing. J Comput Biol 19(5):455–477. https://doi.org/10.1089/cmb.2012.0021
46. Sangwan N, Xia F, Gilbert JA (2016) Recovering complete and draft population genomes from metagenome datasets. Microbiome 4:8. https://doi.org/10.1186/s40168-016-0154-5
47. Frank JA, Pan Y, Tooming-Klunderud A, Eijsink VG, McHardy AC, Nederbragt AJ, Pope PB (2016) Improved metagenome assemblies and taxonomic binning using long-read circular consensus sequence data. Sci Rep 6:25373. https://doi.org/10.1038/srep25373
48. Kang DD, Froula J, Egan R, Wang Z (2015) MetaBAT, an efficient tool for accurately reconstructing single genomes from complex microbial communities. PeerJ 3:e1165. https://doi.org/10.7717/peerj.1165
49. Wu Y-W, Tang Y-H, Tringe SG, Simmons BA, Singer SW (2014) MaxBin: an automated binning method to recover individual genomes from metagenomes using an expectation-maximization algorithm. Microbiome 2(1):26. https://doi.org/10.1186/2049-2618-2-26
50. Alneberg J, Bjarnason BS, de Bruijn I, Schirmer M, Quick J, Ijaz UZ, Lahti L, Loman NJ, Andersson AF, Quince C (2014) Binning metagenomic contigs by coverage and composition. Nat Methods 11(11):1144–1146. https://doi.org/10.1038/nmeth.3103
51. Gregor I, Dröge J, Schirmer M, Quince C, McHardy AC (2016) PhyloPythiaS+: a self-training method for the rapid reconstruction of low-ranking taxonomic bins from metagenomes. PeerJ 4:e1603. https://doi.org/10.7717/peerj.1603
52. Huson DH, Auch AF, Qi J, Schuster SC (2007) MEGAN analysis of metagenomic data. Genome Res 17(3):377–386. https://doi.org/10.1101/gr.5969107
53. Dröge J, Gregor I, McHardy AC (2015) Taxator-tk: precise taxonomic assignment of metagenomes by fast approximation of evolutionary neighborhoods. Bioinformatics 31(6):817–824. https://doi.org/10.1093/bioinformatics/btu745
54. McHardy AC, Rigoutsos I (2007) What's in the mix: phylogenetic classification of metagenome sequence samples. Curr Opin Microbiol 10(5):499–503. https://doi.org/10.1016/j.mib.2007.08.004

55. Sczyrba A, Hofmann P, Belmann P, Koslicki D, Janssen S, Droege J, Gregor I, Majda S, Fiedler J, Dahms E, Bremges A, Fritz A, Garrido-Oter R, Sparholt Jorgensen T, Shapiro N, Blood PD, Gurevich A, Bai Y, Turaev D, DeMaere MZ, Chikhi R, Nagarajan N, Quince C, Hestbjerg Hansen L, Sorensen SJ, Chia BKH, Denis B, Froula JL, Wang Z, Egan R, Kang DD, Cook JJ, Deltel C, Beckstette M, Lemaitre C, Peterlongo P, Rizk G, Lavenier D, Wu Y-W, Singer SW, Jain C, Strous M, Klingenberg H, Meinicke P, Barton M, Lingner T, Lin H-H, Liao Y-C, Gueiros Z, Silva G, Cuevas DA, Edwards RA, Saha S, Piro VC, Renard BY, Pop M, Klenk H-P, Goeker M, Kyrpides N, Woyke T, Vorholt JA, Schulze-Lefert P, Rubin EM, Darling AE, Rattei T, McHardy AC (2017) Critical Assessment of Metagenome Interpretation – a benchmark of computational metagenomics software. bioRxiv. https://doi.org/10.1101/099127
56. Eren AM, Esen ÖC, Quince C, Vineis JH, Morrison HG, Sogin ML, Delmont TO (2015) Anvi'o: an advanced analysis and visualization platform for 'omics data. PeerJ 3:e1319. https://doi.org/10.7717/peerj.1319
57. Zhu Z, Niu B, Chen J, Wu S, Sun S, Li W (2013) MGAviewer: a desktop visualization tool for analysis of metagenomics alignment data. Bioinformatics 29(1):122–123. https://doi.org/10.1093/bioinformatics/bts567
58. Parks DH, Imelfort M, Skennerton CT, Hugenholtz P, Tyson GW (2015) CheckM: assessing the quality of microbial genomes recovered from isolates, single cells, and metagenomes. Genome Res 25(7):1043–1055. https://doi.org/10.1101/gr.186072.114
59. Rodriguez-R LM, Konstantinidis KT (2016) The enveomics collection: a toolbox for specialized analyses of microbial genomes and metagenomes. PeerJ Preprints 4:e1900v1901. https://doi.org/10.7287/peerj.preprints.1900v1
60. Bowers RM, Kyrpides NC, Stepanauskas R, Harmon-Smith M, Doud D, Reddy TBK, Schulz F, Jarett J, Rivers AR, Eloe-Fadrosh EA, Tringe SG, Ivanova NN, Copeland A, Clum A, Becraft ED, Malmstrom RR, Birren B, Podar M, Bork P, Weinstock GM, Garrity GM, Dodsworth JA, Yooseph S, Sutton G, Glockner FO, Gilbert JA, Nelson WC, Hallam SJ, Jungbluth SP, Ettema TJG, Tighe S, Konstantinidis KT, Liu W-T, Baker BJ, Rattei T, Eisen JA, Hedlund B, McMahon KD, Fierer N, Knight R, Finn R, Cochrane G, Karsch-Mizrachi I, Tyson GW, Rinke C, The Genome Standards C, Lapidus A, Meyer F, Yilmaz P, Parks DH, Eren AM, Schriml L, Banfield JF, Hugenholtz P, Woyke T (2017) Minimum information about a single amplified genome (MISAG) and a metagenome-assembled genome (MIMAG) of bacteria and archaea. Nat Biotechnol 35(8):725–731. https://doi.org/10.1038/nbt.3893
61. Trimble WL, Keegan KP, D'Souza M, Wilke A, Wilkening J, Gilbert J, Meyer F (2012) Short-read reading-frame predictors are not created equal: sequence error causes loss of signal. BMC Bioinformatics 13:183. https://doi.org/10.1186/1471-2105-13-183
62. Zhu W, Lomsadze A, Borodovsky M (2010) Ab initio gene identification in metagenomic sequences. Nucleic Acids Res 38(12):e132–e132. https://doi.org/10.1093/nar/gkq275
63. Frank JA, Arntzen MØ, Sun L, Hagen LH, McHardy AC, Horn SJ, Eijsink VGH, Schnürer A, Pope PB (2016) Novel syntrophic populations dominate an ammonia-tolerant methanogenic microbiome. mSystems 1(5). https://doi.org/10.1128/mSystems.00092-16
64. Tang H, Li S, Ye Y (2016) A graph-centric approach for metagenome-guided peptide and protein identification in metaproteomics. PLoS Comput Biol 12(12):e1005224. https://doi.org/10.1371/journal.pcbi.1005224
65. Weimann A, Mooren K, Frank J, Pope PB, Bremges A, McHardy AC (2016) From genomes to phenotypes: traitar, the microbial trait analyzer. mSystems 1(6). https://doi.org/10.1128/mSystems.00101-16
66. Muth T, Benndorf D, Reichl U, Rapp E, Martens L (2013) Searching for a needle in a stack of needles: challenges in metaproteomics data analysis. Mol Biosyst 9(4):578–585. https://doi.org/10.1039/c2mb25415h
67. Nesvizhskii AI, Aebersold R (2005) Interpretation of shotgun proteomic data: the protein inference problem. Mol Cell Proteomics 4(10):1419–1440. https://doi.org/10.1074/mcp.R500012-MCP200

68. Perkins DN, Pappin DJ, Creasy DM, Cottrell JS (1999) Probability-based protein identification by searching sequence databases using mass spectrometry data. Electrophoresis 20(18):3551–3567
69. Cox J, Neuhauser N, Michalski A, Scheltema RA, Olsen JV, Mann M (2011) Andromeda: a peptide search engine integrated into the MaxQuant environment. J Proteome Res 10(4):1794–1805. https://doi.org/10.1021/pr101065j
70. Craig R, Beavis RC (2003) A method for reducing the time required to match protein sequences with tandem mass spectra. Rapid Commun Mass Spectrom 17(20):2310–2316. https://doi.org/10.1002/rcm.1198
71. Eng JK, Searle BC, Clauser KR, Tabb DL (2011) A face in the crowd: recognizing peptides through database search. Mol Cell Proteomics 10(11):R111.009522. doi:https://doi.org/10.1074/mcp.R111.009522
72. Vaudel M, Burkhart JM, Sickmann A, Martens L, Zahedi RP (2011) Peptide identification quality control. Proteomics 11(10):2105–2114. https://doi.org/10.1002/pmic.201000704
73. Elias JE, Gygi SP (2007) Target-decoy search strategy for increased confidence in large-scale protein identifications by mass spectrometry. Nat Methods 4(3):207–214. https://doi.org/10.1038/nmeth1019
74. Jagtap P, Goslinga J, Kooren JA, McGowan T, Wroblewski MS, Seymour SL, Griffin TJ (2013) A two-step database search method improves sensitivity in peptide sequence matches for metaproteomics and proteogenomics studies. Proteomics 13(8):1352–1357. https://doi.org/10.1002/pmic.201200352
75. Chatterjee S, Stupp GS, Park SK, Ducom JC, Yates JR 3rd, Su AI, Wolan DW (2016) A comprehensive and scalable database search system for metaproteomics. BMC Genomics 17(1):642. https://doi.org/10.1186/s12864-016-2855-3
76. Wang Y, Ahn TH, Li Z, Pan C (2013) Sipros/ProRata: a versatile informatics system for quantitative community proteomics. Bioinformatics 29(16):2064–2065. https://doi.org/10.1093/bioinformatics/btt329
77. Gonnelli G, Stock M, Verwaeren J, Maddelein D, De Baets B, Martens L, Degroeve S (2015) A decoy-free approach to the identification of peptides. J Proteome Res 14(4):1792–1798. https://doi.org/10.1021/pr501164r
78. Shevchenko A, Sunyaev S, Loboda A, Shevchenko A, Bork P, Ens W, Standing KG (2001) Charting the proteomes of organisms with unsequenced genomes by MALDI-quadrupole time-of-flight mass spectrometry and BLAST homology searching. Anal Chem 73(9):1917–1926
79. Pevtsov S, Fedulova I, Mirzaei H, Buck C, Zhang X (2006) Performance evaluation of existing de novo sequencing algorithms. J Proteome Res 5(11):3018–3028. https://doi.org/10.1021/pr060222h
80. Frank A, Pevzner P (2005) PepNovo: de novo peptide sequencing via probabilistic network modeling. Anal Chem 77(4):964–973
81. Ma B, Zhang K, Hendrie C, Liang C, Li M, Doherty-Kirby A, Lajoie G (2003) PEAKS: powerful software for peptide de novo sequencing by tandem mass spectrometry. Rapid Commun Mass Spectrom 17(20):2337–2342. https://doi.org/10.1002/rcm.1196
82. Allmer J (2011) Algorithms for the de novo sequencing of peptides from tandem mass spectra. Expert Rev Proteomics 8(5):645–657. https://doi.org/10.1586/epr.11.54
83. Muth T, Kolmeder CA, Salojarvi J, Keskitalo S, Varjosalo M, Verdam FJ, Rensen SS, Reichl U, de Vos WM, Rapp E, Martens L (2015) Navigating through metaproteomics data: a logbook of database searching. Proteomics 15(20):3439–3453. https://doi.org/10.1002/pmic.201400560
84. Craig R, Cortens JC, Fenyo D, Beavis RC (2006) Using annotated peptide mass spectrum libraries for protein identification. J Proteome Res 5(8):1843–1849. https://doi.org/10.1021/pr0602085
85. Lam H, Deutsch EW, Eddes JS, Eng JK, King N, Stein SE, Aebersold R (2007) Development and validation of a spectral library searching method for peptide identification from MS/MS. Proteomics 7(5):655–667. https://doi.org/10.1002/pmic.200600625

86. Frewen B, MacCoss MJ (2007) Using BiblioSpec for creating and searching tandem MS peptide libraries. Curr Protoc Bioinformatics Chapter 13:Unit 13 17. doi:https://doi.org/10.1002/0471250953.bi1307s20
87. Liu H, Sadygov RG, Yates JR 3rd (2004) A model for random sampling and estimation of relative protein abundance in shotgun proteomics. Anal Chem 76(14):4193–4201. https://doi.org/10.1021/ac0498563
88. Ishihama Y, Oda Y, Tabata T, Sato T, Nagasu T, Rappsilber J, Mann M (2005) Exponentially modified protein abundance index (emPAI) for estimation of absolute protein amount in proteomics by the number of sequenced peptides per protein. Mol Cell Proteomics 4(9):1265–1272. https://doi.org/10.1074/mcp.M500061-MCP200
89. Paoletti AC, Parmely TJ, Tomomori-Sato C, Sato S, Zhu D, Conaway RC, Conaway JW, Florens L, Washburn MP (2006) Quantitative proteomic analysis of distinct mammalian Mediator complexes using normalized spectral abundance factors. Proc Natl Acad Sci U S A 103(50):18928–18933. https://doi.org/10.1073/pnas.0606379103
90. Nahnsen S, Bielow C, Reinert K, Kohlbacher O (2013) Tools for label-free peptide quantification. Mol Cell Proteomics 12(3):549–556. https://doi.org/10.1074/mcp.R112.025163
91. Muth T, Behne A, Heyer R, Kohrs F, Benndorf D, Hoffmann M, Lehteva M, Reichl U, Martens L, Rapp E (2015) The MetaProteomeAnalyzer: a powerful open-source software suite for metaproteomics data analysis and interpretation. J Proteome Res 14(3):1557–1565. https://doi.org/10.1021/pr501246w
92. Cox J, Mann M (2008) MaxQuant enables high peptide identification rates, individualized p.p.b.-range mass accuracies and proteome-wide protein quantification. Nat Biotechnol 26(12):1367–1372. https://doi.org/10.1038/nbt.1511
93. Jagtap PD, Blakely A, Murray K, Stewart S, Kooren J, Johnson JE, Rhodus NL, Rudney J, Griffin TJ (2015) Metaproteomic analysis using the Galaxy framework. Proteomics 15(20):3553–3565. https://doi.org/10.1002/pmic.201500074
94. Argentini A, Goeminne LJ, Verheggen K, Hulstaert N, Staes A, Clement L, Martens L (2016) moFF: a robust and automated approach to extract peptide ion intensities. Nat Methods 13(12):964–966. https://doi.org/10.1038/nmeth.4075
95. Huson DH, Beier S, Flade I, Gorska A, El-Hadidi M, Mitra S, Ruscheweyh HJ, Tappu R (2016) MEGAN community edition – interactive exploration and analysis of large-scale microbiome sequencing data. PLoS Comput Biol 12(6):e1004957. https://doi.org/10.1371/journal.pcbi.1004957
96. Mesuere B, Devreese B, Debyser G, Aerts M, Vandamme P, Dawyndt P (2012) Unipept: tryptic peptide-based biodiversity analysis of metaproteome samples. J Proteome Res 11(12):5773–5780. https://doi.org/10.1021/pr300576s
97. Apweiler R, Bairoch A, Wu CH, Barker WC, Boeckmann B, Ferro S, Gasteiger E, Huang H, Lopez R, Magrane M, Martin MJ, Natale DA, O'Donovan C, Redaschi N, Yeh LS (2004) UniProt: the universal protein knowledgebase. Nucleic Acids Res 32(Database issue):D115–D119. https://doi.org/10.1093/nar/gkh131
98. Schneider T, Schmid E, de Castro JV Jr, Cardinale M, Eberl L, Grube M, Berg G, Riedel K (2011) Structure and function of the symbiosis partners of the lung lichen (*Lobaria pulmonaria* L. Hoffm.) analyzed by metaproteomics. Proteomics 11(13):2752–2756. https://doi.org/10.1002/pmic.201000679
99. Penzlin A, Lindner MS, Doellinger J, Dabrowski PW, Nitsche A, Renard BY (2014) Pipasic: similarity and expression correction for strain-level identification and quantification in metaproteomics. Bioinformatics 30(12):i149–i156. https://doi.org/10.1093/bioinformatics/btu267
100. Heyer R, Schallert K, Zoun R, Becher B, Saake G, Benndorf D (2017) Challenges and perspectives of metaproteomic data analysis. J Biotechnol 261:24. https://doi.org/10.1016/j.jbiotec.2017.06.1201
101. Tyanova S, Temu T, Sinitcyn P, Carlson A, Hein MY, Geiger T, Mann M, Cox J (2016) The Perseus computational platform for comprehensive analysis of (prote)omics data. Nat Methods 13(9):731–740. https://doi.org/10.1038/nmeth.3901

102. Luo W, Brouwer C (2013) Pathview: an R/Bioconductor package for pathway-based data integration and visualization. Bioinformatics 29(14):1830–1831. https://doi.org/10.1093/bioinformatics/btt285
103. Altschul SF, Gish W, Miller W, Myers EW, Lipman DJ (1990) Basic local alignment search tool. J Mol Biol 215(3):403–410. https://doi.org/10.1016/S0022-2836(05)80360-2
104. Finn RD, Clements J, Eddy SR (2011) HMMER web server: interactive sequence similarity searching. Nucleic Acids Res 39(Web Server issue):W29–W37. https://doi.org/10.1093/nar/gkr367
105. Prakash T, Taylor TD (2012) Functional assignment of metagenomic data: challenges and applications. Brief Bioinform 13(6):711–727. https://doi.org/10.1093/bib/bbs033
106. Overbeek R, Fonstein M, D'Souza M, Pusch GD, Maltsev N (1999) The use of gene clusters to infer functional coupling. Proc Natl Acad Sci U S A 96(6):2896–2901
107. Tatusov RL, Koonin EV, Lipman DJ (1997) A genomic perspective on protein families. Science 278(5338):631–637
108. Finn RD, Coggill P, Eberhardt RY, Eddy SR, Mistry J, Mitchell AL, Potter SC, Punta M, Qureshi M, Sangrador-Vegas A, Salazar GA, Tate J, Bateman A (2016) The Pfam protein families database: towards a more sustainable future. Nucleic Acids Res 44(D1):D279–D285. https://doi.org/10.1093/nar/gkv1344
109. Haft DH, Selengut JD, White O (2003) The TIGRFAMs database of protein families. Nucleic Acids Res 31(1):371–373
110. Bairoch A (2000) The ENZYME database in 2000. Nucleic Acids Res 28(1):304–305
111. Kanehisa M, Goto S (2000) KEGG: kyoto encyclopedia of genes and genomes. Nucleic Acids Res 28(1):27–30
112. Krieger CJ, Zhang P, Mueller LA, Wang A, Paley S, Arnaud M, Pick J, Rhee SY, Karp PD (2004) MetaCyc: a multiorganism database of metabolic pathways and enzymes. Nucleic Acids Res 32(Database issue):D438–D442. https://doi.org/10.1093/nar/gkh100
113. Ashburner M, Ball CA, Blake JA, Botstein D, Butler H, Cherry JM, Davis AP, Dolinski K, Dwight SS, Eppig JT, Harris MA, Hill DP, Issel-Tarver L, Kasarskis A, Lewis S, Matese JC, Richardson JE, Ringwald M, Rubin GM, Sherlock G (2000) Gene ontology: tool for the unification of biology. The Gene Ontology Consortium. Nat Genet 25(1):25–29. https://doi.org/10.1038/75556
114. Subramanian A, Tamayo P, Mootha VK, Mukherjee S, Ebert BL, Gillette MA, Paulovich A, Pomeroy SL, Golub TR, Lander ES, Mesirov JP (2005) Gene set enrichment analysis: a knowledge-based approach for interpreting genome-wide expression profiles. Proc Natl Acad Sci U S A 102(43):15545–15550. https://doi.org/10.1073/pnas.0506580102
115. Dennis G Jr, Sherman BT, Hosack DA, Yang J, Gao W, Lane HC, Lempicki RA (2003) DAVID: database for annotation, visualization, and integrated discovery. Genome Biol 4(5):P3
116. Reimand J, Kull M, Peterson H, Hansen J, Vilo J (2007) g:Profiler – a web-based toolset for functional profiling of gene lists from large-scale experiments. Nucleic Acids Res 35(Web Server issue):W193–W200. https://doi.org/10.1093/nar/gkm226
117. Finn RD, Attwood TK, Babbitt PC, Bateman A, Bork P, Bridge AJ, Chang HY, Dosztanyi Z, El-Gebali S, Fraser M, Gough J, Haft D, Holliday GL, Huang H, Huang X, Letunic I, Lopez R, Lu S, Marchler-Bauer A, Mi H, Mistry J, Natale DA, Necci M, Nuka G, Orengo CA, Park Y, Pesseat S, Piovesan D, Potter SC, Rawlings ND, Redaschi N, Richardson L, Rivoire C, Sangrador-Vegas A, Sigrist C, Sillitoe I, Smithers B, Squizzato S, Sutton G, Thanki N, Thomas PD, Tosatto SC, Wu CH, Xenarios I, Yeh LS, Young SY, Mitchell AL (2017) InterPro in 2017-beyond protein family and domain annotations. Nucleic Acids Res 45(D1):D190–D199. https://doi.org/10.1093/nar/gkw1107
118. Quevillon E, Silventoinen V, Pillai S, Harte N, Mulder N, Apweiler R, Lopez R (2005) InterProScan: protein domains identifier. Nucleic Acids Res 33(Web Server issue):W116–W120. https://doi.org/10.1093/nar/gki442
119. Markowitz VM, Chen IM, Palaniappan K, Chu K, Szeto E, Grechkin Y, Ratner A, Jacob B, Huang J, Williams P, Huntemann M, Anderson I, Mavromatis K, Ivanova NN, Kyrpides NC

(2012) IMG: the Integrated Microbial Genomes database and comparative analysis system. Nucleic Acids Res 40(Database issue):D115–D122. https://doi.org/10.1093/nar/gkr1044

120. Kanehisa M, Sato Y, Morishima K (2016) BlastKOALA and GhostKOALA: KEGG tools for functional characterization of genome and metagenome sequences. J Mol Biol 428(4): 726–731. https://doi.org/10.1016/j.jmb.2015.11.006

121. Lombard V, Golaconda Ramulu H, Drula E, Coutinho PM, Henrissat B (2014) The carbohydrate-active enzymes database (CAZy) in 2013. Nucleic Acids Res 42(Database issue): D490–D495. https://doi.org/10.1093/nar/gkt1178

122. Yin Y, Mao X, Yang J, Chen X, Mao F, Xu Y (2012) dbCAN: a web resource for automated carbohydrate-active enzyme annotation. Nucleic Acids Res 40(Web Server issue):W445–W451. https://doi.org/10.1093/nar/gks479

123. Park BH, Karpinets TV, Syed MH, Leuze MR, Uberbacher EC (2010) CAZymes Analysis Toolkit (CAT): web service for searching and analyzing carbohydrate-active enzymes in a newly sequenced organism using CAZy database. Glycobiology 20(12):1574–1584. https://doi.org/10.1093/glycob/cwq106

124. Rawlings ND, Barrett AJ, Finn R (2016) Twenty years of the MEROPS database of proteolytic enzymes, their substrates and inhibitors. Nucleic Acids Res 44(D1):D343–D350. https://doi.org/10.1093/nar/gkv1118

125. Huberts DH, van der Klei IJ (2010) Moonlighting proteins: an intriguing mode of multitasking. Biochim Biophys Acta 1803(4):520–525. https://doi.org/10.1016/j.bbamcr.2010.01.022

126. Mayers MD, Moon C, Stupp GS, Su AI, Wolan DW (2017) Quantitative metaproteomics and activity-based probe enrichment reveals significant alterations in protein expression from a mouse model of inflammatory bowel disease. J Proteome Res 16(2):1014–1026. https://doi.org/10.1021/acs.jproteome.6b00938

127. Jehmlich N, Schmidt F, Taubert M, Seifert J, Bastida F, von Bergen M, Richnow HH, Vogt C (2010) Protein-based stable isotope probing. Nat Protoc 5(12):1957–1966. https://doi.org/10.1038/nprot.2010.166

128. Campanaro S, Treu L, Kougias PG, Francisci D, Valle G, Angelidaki I (2016) Metagenomic analysis and functional characterization of the biogas microbiome using high throughput shotgun sequencing and a novel binning strategy. Biotechnol Biofuels 9(1):1

129. Zhou Y, Pope PB, Li S, Wen B, Tan F, Cheng S, Chen J, Yang J, Liu F, Lei X (2014) Omics-based interpretation of synergism in a soil-derived cellulose-degrading microbial community. Sci Rep 4:5288. https://doi.org/10.1038/srep05288

130. Hartmann H, Ahring BK (2005) Anaerobic digestion of the organic fraction of municipal solid waste: influence of co-digestion with manure. Water Res 39(8):1543–1552. https://doi.org/10.1016/j.watres.2005.02.001

131. Westerholm M, Moestedt J, Schnürer A (2016) Biogas production through syntrophic acetate oxidation and deliberate operating strategies for improved digester performance. Appl Energy 179:124–135

132. Gallert C, Winter J (1997) Mesophilic and thermophilic anaerobic digestion of source-sorted organic wastes: effect of ammonia on glucose degradation and methane production. Appl Microbiol Biotechnol 48(3):405–410

133. McInerney MJ, Struchtemeyer CG, Sieber J, Mouttaki H, Stams AJM, Schink B, Rohlin L, Gunsalus RP (2008) Physiology, ecology, phylogeny, and genomics of microorganisms capable of syntrophic metabolism. In: Wiegel J, Maier RJ, Adams MWW (eds) Incredible anaerobes: from physiology to genomics to fuels, Annals of the New York Academy of Sciences, vol 1125. Blackwell Publishing, Oxford, pp 58–72. https://doi.org/10.1196/annals.1419.005

134. Westerholm M, Roos S, Schnürer A (2011) *Tepidanaerobacter acetatoxydans* sp. nov., an anaerobic, syntrophic acetate-oxidizing bacterium isolated from two ammonium-enriched mesophilic methanogenic processes. Syst Appl Microbiol 34(4):260–266

135. Schnürer A, Schink B, Svensson BH (1996) Clostridium ultunense sp. nov., a mesophilic bacterium oxidizing acetate in syntrophic association with a hydrogenotrophic methanogenic bacterium. Int J Syst Bacteriol 46(4):1145–1152

136. Hattori S, Kamagata Y, Hanada S, Shoun H (2000) *Thermacetogenium phaeum* gen. nov., sp. nov., a strictly anaerobic, thermophilic, syntrophic acetate-oxidizing bacterium. Int J Syst Evol Microbiol 50(4):1601–1609
137. Llewellyn MS, Boutin S, Hoseinifar SH, Derome N (2014) Teleost microbiomes: the state of the art in their characterization, manipulation and importance in aquaculture and fisheries. Front Microbiol 5:207. https://doi.org/10.3389/fmicb.2014.00207
138. Karlsen C, Ottem KF, Brevik OJ, Davey M, Sorum H, Winther-Larsen HC (2017) The environmental and host-associated bacterial microbiota of Arctic seawater-farmed Atlantic salmon with ulcerative disorders. J Fish Dis 40:1645. https://doi.org/10.1111/jfd.12632
139. Ángeles Esteban M (2012) An overview of the immunological defenses in fish skin. ISRN Immunol 2012:29. https://doi.org/10.5402/2012/853470
140. Martens EC, Chiang HC, Gordon JI (2008) Mucosal glycan foraging enhances fitness and transmission of a saccharolytic human gut bacterial symbiont. Cell Host Microbe 4(5):447–457
141. Roberts SD, Powell MD (2005) The viscosity and glycoprotein biochemistry of salmonid mucus varies with species, salinity and the presence of amoebic gill disease. J Comp Physiol B 175(1):1–11
142. Edgar RC (2010) Search and clustering orders of magnitude faster than BLAST. Bioinformatics 26(19):2460–2461. https://doi.org/10.1093/bioinformatics/btq461
143. Kuczynski J, Stombaugh J, Walters WA, González A, Caporaso JG, Knight R (2012) Using QIIME to analyze 16S rRNA gene sequences from microbial communities. Curr Protoc Microbiol 1E. 5.1–1E. 5.20
144. Cordero H, Morcillo P, Cuesta A, Brinchmann MF, Esteban MA (2016) Differential proteome profile of skin mucus of gilthead seabream (*Sparus aurata*) after probiotic intake and/or overcrowding stress. J Proteomics 132:41–50. https://doi.org/10.1016/j.jprot.2015.11.017
145. Jurado J, Fuentes-Almagro CA, Guardiola FA, Cuesta A, Esteban MA, Prieto-Alamo MJ (2015) Proteomic profile of the skin mucus of farmed gilthead seabream (*Sparus aurata*). J Proteomics 120:21–34. https://doi.org/10.1016/j.jprot.2015.02.019

Index

© Springer Nature Switzerland AG 2019
J.-L. Capelo-Martínez (ed.), *Emerging Sample Treatments in Proteomics*, Advances
in Experimental Medicine and Biology 1073, https://doi.org/10.1007/978-3-030-12298-0

Printed in the United States
By Bookmasters